Ultra-Low Power Wireless Technologies for Sensor Networks

SERIES ON INTEGRATED CIRCUITS AND SYSTEMS

Anantha Chandrakasan, Editor
Massachusetts Institute of Technology
Cambridge, Massachusetts, USA

Published books in the series:

Ultra-Low Power Wireless Technologies for Sensor Networks
Brian Otis and Jan Rabaey
2007, ISBN 978-0-387-30930-9

Sub-threshold Design for Ultra Low-Power Systems
Alice Wang, Benton H. Calhoun, and Anantha Chandrakasan
2006, ISBN 0-387-33515-3

High Performance Energy Efficient Microprocessor Design
Vojin Oklibdzija and Ram Krishnamurthy (Eds.)
2006, ISBN 0-387-28594-6

Abstraction Refinement for Large Scale Model Checking
Chao Wang, Gary D. Hachtel, and Fabio Somenzi
2006, ISBN 0-387-28594-6

A Practical Introduction to PSL
Cindy Eisner and Dana Fisman
2006, ISBN 0-387-35313-5

Thermal and Power Management of Integrated Circuits
Arman Vassighi and Manoj Sachdev
2006, ISBN 0-398-25762-4

Leakage in Nanometer CMOS Technologies
Siva G. Narendra and Anantha Chandrakasan
2005, ISBN 0-387-25737-3

Statistical Analysis and Optimization for VLSI: Timing and Power
Ashish Srivastava, Dennis Sylvester and David Blaauw
2005, ISBN 0-387-26049-8

Brian Otis and Jan Rabaey

Ultra-Low Power Wireless Technologies for Sensor Networks

 Springer

Brian Otis
University of Washington
Department of Electrical Engineering
Seattle, WA, USA

Jan Rabaey
University of California, Berkeley
Department of Electrical Engineering and Computer Science
Berkeley, CA, USA

Ultra-Low Power Wireless Technologies for Sensor Networks

Library of Congress Control Number: 2006936495

ISBN 0-387-30930-6   e-ISBN 0-387-49313-1
ISBN 978-0-387-30930-9   e-ISBN 978-0-387-49313-8

Printed on acid-free paper.

9 8 7 6 5 4 3 2 1

springer.com

Table of Contents

List of Figures

List of Tables

1

WIRELESS SENSOR NETWORKS

The ubiquity of wireless links in our everyday life is increasing dramatically. One example of this is the emerging field of wireless sensor networks. This new field presents many opportunities and challenges. One particularly difficult aspect of wireless sensing is the implementation of the radio link.

This book is aimed at academic and professional engineers designing communication links for pervasive and low power applications. New technologies are introduced, design techniques are given, and theoretical limits are described. Design philosophies and experimental results from four leading sensor network hardware groups are presented. The topics presented here can be applied to sensor network transceiver design and extended to all low power, highly integrated radio applications.

The accurate quantification of environmental parameters has been an important part of human existence for centuries. In 1612, Santorio Santorio invented the graduated thermometer, and human curiosity and quest for knowledge has been growing ever since. Modern sensing parameters include humidity, seismic activity, building occupation, airflow, particulate detection, and countless medical and biological sensing parameters. A *wireless sensor* monitors one or more environmental parameters and transmits the data to the network for automatic control or human evaluation. This process could include data compression and accumulation to reduce the amount of transmitted data. One could conceive of a huge network of these sensors that are able to communicate with each other. Decades of technological advances have made it conceivable to build and deploy dense wireless networks of autonomous nodes collecting and disseminating wide ranges of environmental data. Wireless sensor networks have many valuable potential applications, ranging from closed-loop environmental control of office buildings and homes to wildlife monitoring. Other possible applications are robot control and guidance in automatic manufacturing environments, warehouse inventory, integrated patient monitoring, diagnostics and drug administration in hospitals, home providing security, identification and personalization, and interactive museums. Successful deployment of wireless sensor and actuator networks in sufficient numbers to provide true ambient intelligence will require the confluence of progress in several disciplines: networking, low power RF and digital IC design, MEMS techniques, energy scavenging, and packaging. Figure 1.1 shows the various specialized blocks of a sensor node.

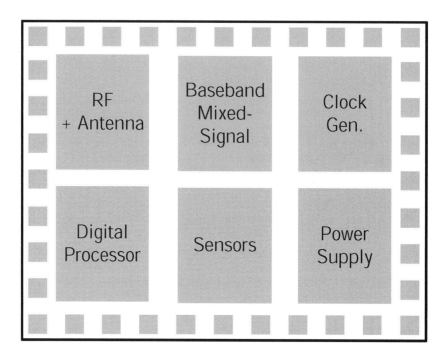

Fig. 1.1. Hardware blocks for sub-mW sensor network implementations

In large-scale system designs, some of these blocks (clock generation, for example) are often considered negligible in terms of power and volume. However, in the implementation of extremely small sensor nodes, each of these components becomes crucial. The most challenging aspect of sensor node implementation is the integration of ultra-low power RF transceivers. It will be shown that the requirements of these transceivers are fundamentally different from those of existing wireless applications, such as cellular telephony, Bluetooth specification radios, and wireless local area network (WLAN) transceivers. The goal of this book is to address some of the design challenges involved with integrating RF communication functionality in a self-contained sensor node.

This chapter introduces the University of California, Berkeley PicoRadio project and describes system level requirements for ad-hoc sensor nodes. The RF transceiver challenges are discussed, including range, sensitivity, power consumption, and integration requirements. Additionally, we introduce several invited chapters to demonstrate the diverse body of work in this field.

1.1 System Requirements

The *PicoRadio* Project at the University of California, Berkeley was founded by Professor Jan Rabaey to identify and address technical barriers to the implementation and large-scale deployment of wireless ad-hoc sensor networks [4]. As will be shown in this chapter, the most challenging aspect of the system implementation is the radio frequency communication datalink. We will begin with a description of the system requirements.

To make dense node deployment possible in practical scenarios, each node must be physically and economically unobtrusive. A node's "obtrusiveness" can be measured by three important metrics: energy scavenging capability, cost and size. These requirements are outlined below.

- **Energy Scavenging:** To reduce installation cost and to allow for a flexible method of deployment, most nodes must be untethered. Maintenance cost considerations render frequent replacement of the energy source of the node unrealistic. Thus, the nodes must scavenge their energy from the environment. This leads to a very aggressive node average power budget of approximately 100μW.
- **Low Cost Implementation:** Sensor node!CostCommercial viability dictates that the unit cost ($\frac{\$}{m^2}$) of the wireless sensor network mesh be very small. For network reliability and high sensing resolution, the node density ($\frac{nodes}{m^2}$) must be high. Thus, the cost of each node must be extremely low ($<1\$$US).
- **Small Form Factor:** Embedding the components into the existing infrastructure of daily environments (walls, furniture, lighting, etc.) requires a very small form factor of the entire sensor node. Typically, node volumes less than 1cm^3 (much smaller than a AA battery) are necessary. A very high level of integration is mandatory if such small dimensions are to be achieved.

The ubiquitous deployment of sensor nodes is economically feasible only if the individual nodes are negligible in cost and size. Achieving such a diminutive stature requires a minimal number of components, a high level of integration, simple and cheap packaging and assembly, and avoidance of expensive components and/or technologies. In the next section, energy scavenging for wireless sensor networks will be discussed, which determines the transceiver power consumption requirements.

1.2 Energy Scavenging

The considerations discussed above dictate that each sensor node be self-sufficient from an energy perspective for the lifetime of the node. This may span up to 10 years for certain building environmental control applications. The energy storage capability of a node is limited by the storage medium (battery or capacitor) and the size constraints. While a single-time charge could work for applications with life cycles below one year, replenishment of the energy supply using energy scavenging is often a necessity. Table 1.1 illustrates the finite power density of state-of-the-art energy sources [3].

Table 1.1. Average power density of various energy storage and scavenging devices [3]

Power Source	Power Density $\frac{\mu W}{cm^3}$	Lifetime
Lithium Battery	100	1 year
Micro Fuel Cell	110	1 year
Solar Cell	10-15000 $(\frac{\mu W}{cm^2})$	∞
Vibrational Converter	375	∞
Air Flow	380	∞
Temperature Gradients	50	∞

Thus, the average power dissipation of the node is severely constrained by the energy scavenging volume of the node. These sources can be broadly grouped into two categories: *energy scavenging sources* and *energy storage sources*. From a volume of 1cm^3, an average continuous power output of 100μW could be supplied by one or a combination of these power sources. If a one year lifetime were acceptable, either a lithium battery or fuel cell would suffice. However, micro fuel cell technology is still in the early stages of research, and is prohibitively complex and expensive. Another active area of research is in thin-film battery technology, which will yield large benefits for sensor node implementations. For desired node lifetimes greater than one year, however, 1cm^3 does not provide ample storage for the node's 3110 $\frac{J}{year}$ energy requirements. Typical node deployment scenarios would demand a 10 year lifetime (31kJ). This is a prohibitively large amount of energy to store in a 1cm^3 volume, necessitating the harvesting of energy from the environment.

Solar power is a proven, universal method of collecting ambient energy. For outdoor or high-light conditions, this is the obvious solution. However, in dim lighting conditions, the power output drops dramatically. In these environments, an additional energy source is needed. Vibrational converters, air flow generators, and temperature gradient generators all produce 50-400 $\frac{\mu W}{cm^3}$, as listed in Table 1.1. Of the three, vibrational converters are the simplest and have the most potential for wafer-scale fabrication. In conclusion, a 1cm^3 sensor node can support an average power draw of 100μW. A combination of solar and vibrational energy scavenging and battery energy storage is likely to yield the most robust and inexpensive solution.

In addition to limitations on average power dissipation, the available peak power levels that can be supplied to the electronics are also limited. Since most microscale energy scavenging and storage devices provide a naturally high impedance, the peak current drive capability is small (less than a few mA). Providing high drive current would require excessively large storage capacitors and complex voltage regulators. The RF datalink circuit design must address this issue by presenting a low peak active current draw.

The next section describes the implication of the energy scavenging requirements on the RF datalink.

1.3 RF Transceiver Requirements

This section describes the transceiver requirements that are unique to sensor node communication. As will be shown, the radio requirements are very different from traditional low power transceivers (pager receivers, RFID tags, Bluetooth-specification radios, keyless-entry).

1.3.1 Power Consumption

In the design of prototype sensor nodes, the wireless interface consumes the largest fraction of the power and size budget of the node. While the demands of the sensing and digital processing components cannot be ignored, their duty cycle is typically very low. A combination of advanced sleep, power-down, and leakage reduction techniques makes it possible to make their average power dissipation virtually negligible [5]. Thus, the wireless interface for sensor networks is the dominant source of power consumption. While optical communication approaches offer the potential of very low power and small size, line-of-sight and directivity considerations make them less attractive [6]. For this work, we will limit our discussion to RF interfaces.

1.3.2 Datarate

As mentioned, the requirements of a transceiver for wireless sensor networks differ dramatically from a traditional wireless link. Thus, common performance metrics such as $\frac{energy}{bit}$ and $\frac{bits}{Hz}$ should be applied with the realization that other factors prevail. For example, a modified metric such as $\frac{energy}{useful-bit}$ is relevant if all sources of power and overhead (for example: synchronization, the impact on energy storage) are included. First we will examine the typical operation mode of the sensor node. An investigation of the traffic patterns and data payloads reveal that the transceiver operation is fundamentally different than a wireless LAN or Bluetooth-specification radio. Data packets in sensor networks tend to be relatively rare and unpredictable events. In most application scenarios, each node in the network sees only a few packets/second. In addition, the packets are relatively short (typically less than 200 bits/packet). This is expected as the payloads normally represent slowly varying and highly correlated environmental data measurements. Combined, this means that the average data rate of a single node rarely exceeds 1 $\frac{kbit}{s}$. These observations are of foremost importance when designing the wireless transceiver, as we will highlight in the following sections.

1.3.3 Range

In the rest of the discussion, we will assume that the sensor networks of interest are dense, which means that the nodes in the network are placed relatively closely (the average distance between nodes is less than or equal to 10m). For a given sensitivity, scaling the node to larger ranges would require additional transmit power or increased coding gain (longer transmit times). As the transmitted power increases in low power transmitters, the global transmitter efficiency increases. Thus, in short-distance links, increasing the transmitted power the preferred approach over increased coding gain. As the transmitted power increases, a linear increase in the link budget is obtained for a sub-linear increase in the transmitter power consumption. Improving the link budget through coding gain would involve linear or super-linear increases in the receive power consumption due to increased packet length and/or higher received bandwidths. Indeed, at transmitted power levels of -10dBm and below, a majority of the transmit mode power is dissipated in the circuitry and not radiated from the antenna. However, at high transmit levels (over 0dBm), the active current draw of the transmitter is high. It is difficult to source high active currents with micro-scale energy scavengers and batteries. Convenient and efficient transmit power levels for sensor node applications are roughly in the range of -10 to +3 dBm.

1.3.4 Sensitivity

Figure 1.2 plots the theoretical range for a radio with a -70dBm sensitivity for various RF propagation models at 2GHz. As shown, the range varies greatly depending on the radio environment. For free space (where the path loss exponent r=2), a range of 37m is achieved with a 0dBm transmit power. However, in indoor environments, a higher exponent (r=3 or r=4) is more appropriate. In that regime, a transmit power of at least 0dBm is required for a 10m range. To add a margin for deep fading, the receiver sensitivity for a 0dBm transmit signal and a 10m range should be greater than -75dBm. Thus, for this application, a receiver sensitivity of better than -75dBm is imposed. Higher sensitivities will allow lower transmitted power levels, subject to the constraints in the previous section.

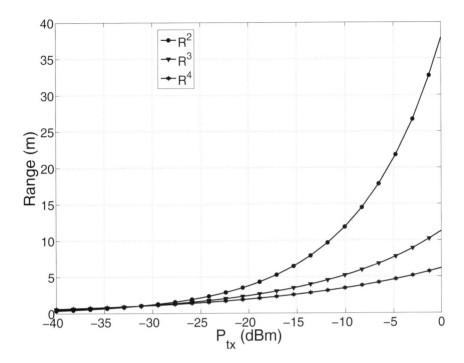

Fig. 1.2. Radio range for receiver with a -70dBm sensitivity

1.3.5 Turn-On Time

In an environment in which the radio is in idle or off mode most of the time, and in which data communications are rare and packets short, it is essential that the radio start up very quickly. For instance, a typical 1Mbps Bluetooth-specification radio with a $500\mu s$ turn-on time would be poorly suited for the transmission of short packets. The on-time to send a 200 bit packet would be only $200\mu s$. Start-up and acquisition represent an overhead that is larger than the actual payload cost, and could easily dominate the power budget (given that channel acquisition is typically the most power-hungry operation). Thus, fast start-up and acquisition is essential to minimize this overhead. An agile radio architecture that allows for quick and efficient channel acquisition and synchronization is desirable. Complex wireless transceivers tend to use sophisticated algorithms such as interference cancellation and large constellation modulation schemes to improve bandwidth efficiency. These techniques translate into complex and lengthy synchronization procedures and may require accurate channel estimations. Packets are spaced almost seconds apart, which is beyond the coherence time of the channel. This means that these procedures have to be repeated for every packet, resulting in major overhead unsuitable in a low-power environment. Simple modulation and communication schemes are hence the desirable solution if agility is a prime requirement.

1.3.6 Integration/Power Tradeoff

Achieving the goal of a very low power/low cost RF design is complicated by a well documented power/integration (cost) tradeoff. For example, the use of high performance SiGe processes, while offering the designer high f_T operation and low bias current levels, eliminates the possibility of integration with low power digital systems. A multi-chip solution would prohibitively increase the cost and area for sensor network applications. Another common strategy for CMOS RF designers trying to reduce power consumption is to use high quality passive surface mount components [7]. This solution also prohibitively increases cost and board area, as each surface mount inductor is larger than the entire transceiver chip.

Recently published "fully integrated" transceivers typically refer to a transceiver that has simply eliminated the need for external ceramic or surface acoustic wave (SAW) filters. They still, however, require an off-chip quartz crystal and various passive components. To meet the cost and form-factor requirements of this application, a truly fully integrated transceiver is mandatory. In addition to increasing the size, off-chip passives add to the complexity and cost of the board manufacturing and package design. In addition, these macro-fabricated components increase the manufactured performance distributions of the radio by adding completely uncorrelated component variations. One method that can be used to achieve a high level of integration is the use of a relatively high carrier frequency. Currently available simple low power radios, as used in control applications, typically operate at low carrier frequencies between 100 and 800MHz. A high carrier frequency has the distinct advantage of reducing the required values of the passive components, making integration easier. For example, a 2.53μH inductance is needed to tune out a 1pF capacitor in a narrow-band system at 100MHz, requiring a surface mount inductor. For a 2GHz carrier frequency, the inductance needed is only 6.33nH, which can easily be integrated on-chip using interconnect metallization layers. In addition, the critical antenna physical dimensions are linearly related to the carrier frequency. For a given antenna radiation pattern and efficiency, a higher carrier frequency allows for a much smaller antenna. A quarter-wavelength monopole antenna at 100MHz would be 0.75m long. At 2GHz, the size shrinks to 37.5mm, allowing very efficient and inexpensive board-trace antennae. However, the drive to higher carrier frequencies in the interest of high integration is in direct conflict with the need for low power consumption. As the carrier frequency increases, the active devices in the RF signal path must be biased at higher cutoff frequencies, increasing the bias current and decreasing the transconductance-to-current $\frac{g_m}{I_d}$ ratio. The results is an increased power dissipation at higher carrier frequencies. Thus, there exists an inherent tradeoff between integration and power consumption that must be addressed through architectural decisions and the use of new technologies.

To reduce the die area and cost of the transceiver, the entire physical layer must be taken into account. For example, a simple, low power front-end with a large, power hungry A/D and complex digital baseband would be counterproductive. Thus, modulation schemes, datarates, and packet sizes that reduce baseband complexity must be embraced to optimize the global power consumption.

In conclusion, RF transceivers for wireless sensor networks must be simple, consume a minimum amount of on-current, and operate at high carrier frequencies to allow high levels of integration. In the next section, the current state-of-the-art in low power RF transceivers is discussed.

There has been substantial work in the field of low power transceivers for wireless sensing. Much of this work has focused on scaling down the power and datarate of traditional transceiver architectures for use in wireless sensing applications. As a case study, a two recent architectures reported in [8] and [9] will be discussed.

The first case study, reported by Peiris [8], achieves impressive sensitivity performance and low power consumption, truly representing the state-of-the-art in heterodyne transceivers for sensor networks.

The receiver, which can utilize either OOK or FSK modulation, was implemented in a $0.18\mu m$ CMOS process. Integration of a the transceiver, A/D converters, RISC microcontroller core and an SRAM block were achieved on the same chip. The receiver consumed 2.1mA from a 1V supply with a high sensitivity of -111dBm and -108dBm in the 433MHz and 868MHz band, respectively. The low current consumption is possible due to the relatively low carrier frequency used. External high Q inductors in the VCO tank reduce the power consumption and improve the phase-noise of the synthesizer block. The VCO phase noise was measured as -110 $\frac{dBc}{Hz}$ at 600kHz offset. The low carrier frequency prohibits the full integration of the VCO and LNA matching inductors, necessitating off-chip components.

The second case study, reported by Molnar [9], uses an FSK modulation scheme with a 900MHz carrier frequency. The transceiver operated from a 3V supply and dissipated 1.2mW for a sensitivity of -93dBm. A low-IF architecture was used. The transceiver made use of the relatively large supply voltage to share bias current between the LNA and the local oscillator. To reduce the power consumption, an off-chip high Q inductor was used for the VCO. This work helped form the foundation for the material presented in Chapter 8.

This book will present efforts from various researchers that further advanced the state-of-the-art in low power RF datalinks for sensor networks.

1.4 Contributions of this Book

This book explores the following three areas:

1. **CMOS/MEMS Co-Design:** Techniques for designing new MEMS technology with sub-micron CMOS circuitry are introduced. Many design examples are presented, including RF local oscillators, RF amplifiers, and low frequency reference clocks. Both low frequency electrostatic silicon resonators and high frequency bulk-acoustic wave resonators are covered. The increasing f_T of sub-micron CMOS processes is leveraged by using weak inversion biasing in all circuit blocks, from DC to 2GHz.

2. **Ultra-Low Power Radio Design:** Techniques and philosophies increasing the level of integration and decreasing the power consumption of CMOS transceivers beyond the current state-of-the-art are presented. A RF transmit beacon operating purely from scavenged energy is presented. Four fully functional receivers have been designed, implemented, and tested to validate these concepts. The first, a two-channel tuned radio frequency (TRF) architecture, demonstrates passive channel selection using MEMS resonators and is scalable to multiple channels. The second, using a super-regenerative architecture, achieves extremely low power dissipation ($400\mu W$) by using a MEMS resonator to set the RF frequency. Third, a low power Ultra Wide Band (UWB) transceiver developed at IMEC, Belgium is presented. Finally, a low voltage narrowband transceiver developed by Ben Cook at U.C. Berkeley is presented.

3. **Integration Techniques for Wireless Sensor Nodes:** To make inexpensive, massively-deployable sensor nodes a reality, extremely small form factors are mandatory. This book explores techniques developed in the PicoRadio project at U.C. Berkeley and at IMEC, Belgium that addresses these issues. For example, the packaging and assembly of systems using MEMS resonators is discussed and advanced, modular packaging of a cubic centimeter system-in-package (SiP) node is presented.

The book will begin with Chapter 2, which outlines design philosophies used in recent low power RF transceiver research. Three techniques are explored: circuits utilizing MEMS/IC co-design, weak-inversion operation of RF circuits, and energy scavenging. To verify these philosophies, three proof-of-concept designs are presented: two RF circuit blocks and one system implementation. Advanced RF MEMS/IC circuit co-design research includes the analysis, design, and implementation of a fully differential oscillator. Section 2.4 describes the circuit and compares the testing results to a single ended version with the same power consumption. To demonstrate the potential for energy scavenging, a 2GHz transmit beacon operating indefinitely on scavenged energy is discussed.

Based on these concepts, a fully functional two-channel transceiver was designed, implemented, and tested. Chapter 3 fully documents this effort from the initial analysis phase through the testing methodology.

To further reduce power consumption, a super-regenerative radio architecture was investigated. A fully functional, prototype $400\mu W$ receiver was demonstrated. Chapter 4 presents the theory behind the receiver operation.

The circuit design, implementation, and subsequent receiver testing results are revealed.

Chapter 5 describes a fully-integrated transceiver based on the super-regenerative radio. The pulse-width demodulator synthesis, analysis, and circuit design is described. In addition, a digitally programmable interface is implemented, which allows control over the receiver biasing, super-regenerative RF frequency, and the transmitter output tank frequency.

Chapter 6 describes an efficient and robust integration methodology for the transceiver. To address an important bottleneck in the implementation of wireless sensor networks, a proof-of-concept CMOS/MEMS 16MHz reference clock with amplitude control loop was designed and implemented. Section 6.1 provides a brief background into micromachined silicon resonators before describing the circuit design. Robust packaging techniques of the MEMS and CMOS components are then discussed. Performing a flip-chip of the MEMS chips directly onto the CMOS chips provides a low parasitic, low form-factor system that saves both die and board area. Two proof-of-concept examples are provided.

Next, a series of invited chapters will reveal interesting design methodologies and techniques used by leading research groups around the world. The first is Chapter 7 by Julien Ryckaert and Steven Sanders from IMEC, Belgium.

Wireless sensors have to demonstrate sufficient energy autonomy in order to make the dream of a seamless ubiquitous environment come through. As being the most practical ressource of power, energy scavenging sources set an arduous limit of about 100μW on the operating power of the radio. Only new approaches in the design of the radio can break the energy gap with today's radios in which most of the power is consumed by just switching on the radio. Therefore, the low duty-cycles and data rates of sensor network applications must be exploited. In this context, ultra-wideband (UWB) technology, in its pulse-based flavor, appears as a very interesting signaling scheme. The short time duration of the useful signal allows turning off the radio between the transmitted and received impulses offering a large potential in the reduction of the radio static power consumption. In the first part of this chapter, a specific UWB air interface is defined in line with a radio architecture in order to provide the low power operation together with a large flexibility in the signal spectrum management as well as high integration capabilities. Using the specification of the air interface definition, a UWB pulse generator fully integrated in a CMOS 0.18μm is discussed in details. Its power consumption of 2mW at 40MHz pulse repetition rate demonstrates the potential of pulse-level duty-cycling for ultra-low power operation. In a second part of the chapter a low-power receiver architecture using a quadrature analog correlation technique is developed for the specific UWB air interface. The system implementation in silicon demonstrates a measured power consumption as low as 16mW for a 20MHz pulse rate. Finally, the last part of the chapter addresses radio integration and packaging aspects as being a critical element in the final cost and size of a sensor device. The chapter concludes with a specific application example of such technology being the wireless body area network applications.

Next, Chapter 8 by Ben Cook and Kris Pister, U.C. Berkeley will explore an ultra-low power narrowband CMOS transceiver. In this chapter, we will explore the energetic requirements of RF wireless communication from both a theoretical and practical standpoint. We focus on energy per transferred bit rather than continuous power consumption because it is more closely tied to the battery life of a wireless device. We begin with a look at the fundamental lower limit on energy per received bit imposed by the celebrated channel capacity theorem set forth by Claude Shannon. Based on this lower bound, we derive an energy efficiency metric for evaluating practical RF systems. By examining power-performance tradeoffs in RF system design, we begin to understand why and by how much will practical systems exceed this fundamental energy bound. From the discussion of system tradeoffs emerge a handful of low energy design techniques allowing systems to move closer to the fundamental energy bound. Finally, measurements of a low energy 2.4GHz transceiver implemented in a 130nm RF CMOS process are discussed.

Finally, Chapter 9 concludes the book with a summary and a discussion of potential directions for future work.

2

LOW POWER CMOS DESIGN FOR RADIO FREQUENCIES

In Chapter 1, an integration/power tradeoff was identified as one main obstacle to the implementation of small, cheap, low power transceivers for wireless sensor networks. This chapter describes the circuit design philosophies that have been developed to address these tradeoffs.

An important metric for RF blocks is the transconductance, which typically determines the total current consumption of the circuit. This chapter begins with a discussion of weak inversion circuit design, which is an increasingly relevant way of achieving efficient transconductance and is becoming more feasible with each technology node. Next, combining MEMS components with CMOS circuitry is shown to greatly reduce the reliance on external passive components. Three proof-of-concept chips are discussed to demonstrate the validity of these design strategies:

1. Circuit Proof-of-Concept I: 300μW Pierce BAW-Oscillator
2. Circuit Proof-of-Concept II: Differential $300\mu W$ BAW-Based Oscillator
3. System Proof-of-Concept: Energy Scavenging Transmit Beacon

We now turn to a discussion of weak inversion CMOS for Radio Frequencies.

2.1 Weak Inversion RF CMOS

As technology scaling relentlessly provides increasing digital clock frequencies, the effects on analog/RF circuit design continue to accumulate. Aggressive supply voltage scaling impedes the implementation of high dynamic range A/D converters and other precision analog circuitry. However, the increasing thickness and reduced resistivity of the copper metallization, coupled with the increased distance of these traces to the lossy substrate, allows the fabrication of high quality inductors and transmission lines. Subthreshold circuit design has been identified as a powerful tool in reducing the power consumption of the transceiver. Weak inversion circuit biasing has been used for years in low frequency analog circuit design [10] to achieve increased transconductance efficiency. However, increasing transistor f_Ts allows weak-inversion operation in some RF circuit blocks. The inversion coefficient (IC) describes the relative level of channel inversion, which is most easily determined by the device current density. As the current density decreases, the transconductance efficiency increases. Equation 2.1 shows the inversion coefficient as a function of the transistor process parameters and operating point.

$$IC = \frac{I_d}{2nk'\frac{W}{L}V_t^2} = \frac{I_d}{I_oS} \qquad (2.1)$$

Where n is the subthreshold slope factor, $k' = \mu C_{ox}$, $V_t = \frac{KT}{q}$, I_o is the specific device current, and S is the transistor aspect ratio. See Figure 2.1 for the transconductance efficiency vs. the inversion coefficient (IC) for a commonly used $0.13\mu m$ process [11].

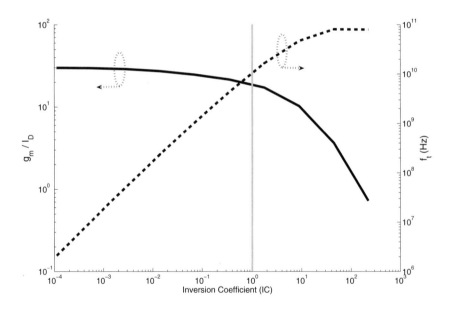

Fig. 2.1. The transconductance efficiency is plotted vs. the inversion coefficient (Courtesy of Nathan Pletcher)

The plot clearly shows the increase in transconductance efficiency as the inversion coefficient decreases. The line at IC=1 is considered the middle of moderate inversion. Since the performance of many RF circuit blocks is directly related to device transconductance, it is advantageous to maximize the transconductance efficiency of all critical devices in order to reduce the necessary bias current. However, reducing the current density also results in a severely decreased device f_T. An optimization of the current density is required to provide the correct balance between transconductance efficiency and bandwidth. Plots such as Figure 2.1 are useful tools for designers when choosing appropriate transistor bias points. Technology scaling allows greatly increased f_T realization for a given IC. Thus, weak inversion biasing for RF design will become increasingly useful in future technology nodes.

Throughout this work, the IC of critical transistors will be discussed. Most of the RF devices are biased in moderate to weak inversion to achieve enhanced transconductance efficiency and reduced bias current.

2.2 MEMS Background

The relatively new field of Radio Frequency Microelectro Mechanical Systems (RF MEMS) provides unique opportunities for RF transceiver designers. This section provides background on RF MEMS and provides insight into the opportunities presented by these new technologies. The field of RF MEMS includes the design and utilization of RF filters, resonators, switches, and other passive mechanical structures constructed using bulk processed integrated circuit fabrication techniques. To date, these devices have been commercially used as discrete board-mounted components, primarily used to enhance the miniaturization of mobile phones. However, RF MEMS components have the potential to be batch fabricated using existing integrated circuit fabrication techniques. Recent capacitively driven and sensed structures offer the potential of integration on the same substrate as the CMOS circuitry. In addition, because the resonant frequency of the structure is set lithographically, rather than by a deposition layer thickness, it is possible to fabricate devices with many unique resonant frequencies on the same die. A good example of this is presented by Bircumshaw in [12]. This resonator was constructed of micromachined polysilicon on a silicon wafer, providing good RF performance and the possibility of integration with active circuitry. The continued improvement in the performance, reliability, and manufacturability of these structures will greatly change the performance and form-factor of RF transceivers by allowing the reliable fabrication of advanced mechanical structures on the same substrate as the circuitry. However, as will be shown in this chapter, these devices hold the potential to enable new circuit blocks and architectures even in their present state as off-chip components.

At the onset of this project, RF MEMS was identified as an emerging technology with the potential to benefit low power RF transceivers. These devices provide three main benefits to circuit designers:

1. **High quality factor (Q) resonant structure.** BAW resonators can achieve quality factors greater than 1000, about 100X higher than on-chip

LC resonators. The resulting increased RF filtering ability both reduces oscillator phase noise and reduces frequency pulling/pushing of oscillators. When used in the design of bandpass filters and duplexers, high Q resonators help to realize the steep skirts necessary to meet cell phone specifications [13]. High Q resonators are further useful in a variety of other transceiver blocks. For example, high Q resonators provide the potential for radio frequency channel select filtering, as their bandwidth is much narrower than what can be obtained from integrated LC filters. This passive channel select filtering can be exploited to simplify the receiver architecture and to reduce the number of active components. In addition, when used in an RF oscillator, RF MEMS resonators provide a vastly improved phase noise compared to a standard, low Q LC resonator [14].

2. **Passive RF frequency reference.** For all narrowband communication systems, an RF carrier frequency generator is necessary. The absolute frequency reference used is typically a low frequency quartz crystal oscillator. The low frequency sinusoid is then multiplied up to radio frequencies by a frequency synthesizer. This technique has a few disadvantages for low power radio design. First, even for a fully integrated frequency synthesizer, an off-chip quartz crystal is always necessary, rendering true full integration impossible. In addition, frequency synthesizers are a large source of power dissipation in low power radios [7]. The VCO and frequency dividers tend to dominate the power consumption of frequency synthesizers. Radio frequency MEMS components provide an inherent high-frequency reference without the need for a power hungry frequency synthesizer.

3. **CMOS/MEMS co-design.** Since MEMS structures are fabricated using the same thin-film fabrication techniques as integrated circuits, each device may be custom-designed for its intended application. This provides additional degrees-of-freedom to the circuit designer. Unlike quartz crystals or discrete 50Ω filters, this flexibility allows the circuit designer to control impedance levels for minimal power consumption. One of the most exciting aspects of RF micromachined components is the potential for co-designing the MEMS devices with the CMOS circuitry. Until now, passive components are either low quality on-chip devices (inductors, capacitors) or high quality off-chip components (inductors, SAW filters, quartz crystals, duplexers). The on-chip components allow customization to meet the requirements of the circuitry, but their performance is usually poor. Meanwhile, high quality off-chip passives offer few designer degrees of freedom. For example, most filters and duplexers are designed for 50Ω input and output impedances. This rigid impedance level is very detrimental from a low power point of view, and has been extremely troublesome in past receiver implementations [15]. The potential of integrating RF MEMS components and circuitry on the same die or on the same substrate using, for instance, fluidic self assembly (FSA) could allow the circuit designer to size the MEMS components and the circuitry simultaneously [16].

The ability to design these devices alongside the circuitry provides increased system performance and additional designer degrees of freedom.

Currently, many industrial and academic institutions have begun development of RF MEMS resonators. One promising technology is the Bulk Acoustic Wave (BAW, FBAR) piezoelectric resonator [17]. Recent work by Aissi has demonstrated the integration of an FBAR resonator on the same substrate as a BiCMOS oscillator [18], possibly foreshadowing the eventual practice of foundries offering MEMS process options. In this section we will provide a description of the structure and electrical model of a bulk acoustic wave resonator.

The FBAR employs a metal-piezo-metal sandwich to achieve a high frequency, tightly controlled second order resonance with an unloaded Q of approximately 1200. As shown in Figure 2.2, the resonator can be modeled as a series LCR circuit, with a series resonance occurring at $f_s = \frac{1}{2\pi\sqrt{L_x C_x}}$, with typically resonant frequencies ranging from 500MHz to 5GHz. Capacitor C_o

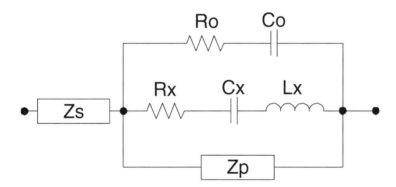

Fig. 2.2. Simplified circuit equivalent model of a BAW resonator

represents a parasitic feedthrough capacitance created by the parallel plates of the resonator. R_o models the finite quality factor of the feedthrough capacitance. Z_s and Z_p represent the loading effects of the CMOS circuitry on the resonator. These external impedances have a large effect on the resonator quality factor, resonant frequency, and frequency stability. In current applications, FBARs are combined into ladder structures and used as duplexers and bandpass filters for wireless applications [17]. Used in this manner, a much smaller form factor is achieved over traditional RF duplexers and filters. For low power transceivers, it is desirable to use single resonators to achieve a high quality factor and very narrow bandwidth. Referring to Figure 2.2, it is possible to distinguish two different resonant modes. The first, occurring at the *series resonance*, allows a low impedance path through the resonator at RF frequencies. At the series resonance, the effective series impedance is approximately

R_x. Above the series resonance, the structure looks inductive and will naturally tune out any parallel capacitive elements. This produces the *parallel resonance*, which occurs approximately 20MHz above the series resonance. At the parallel resonance, there is a circulation of RF current through the resonator and the feedthrough capacitor C_o. Due to this current circulation, at the parallel resonance the impedance of the structure increases dramatically to a value approximately 3 orders of magnitude higher than the *series resonance*. To avoid detuning the resonator in the series resonant mode, the impedance added to the resonator by the electronics (Z_s) must be small compared to R_x, or a few ohms. This is typically not possible using low power RF design techniques, where high impedances are necessary to reduce the current consumption. To avoid detuning the parallel resonant mode, the shunting impedance presented by the electronics (Z_p) must be much higher than the resonator impedance at parallel resonance (approximately 2kΩ). Thus, for low power RF transceiver design, it is desirable to operate on the parallel resonance of the resonator [19]. It should be noted, however, that the intrinsic quality factor of the series resonance will be higher than the parallel resonance due to the additional resistive losses of the circulating RF current.

The frequency stability of high Q resonators is a key feature that provides much more reliable operation that other LC-based frequency generators. Equation 2.2 shows the sensitivity of the parallel resonance[1] to capacitive variation.

$$\frac{\delta f_p}{\delta C_T} \simeq f_{series} \frac{-C_x}{2C_T^2} \qquad (2.2)$$

C_T describes the total capacitive loading on the resonator This equation corresponds to a sensitivity of approximately -10kHz/fF of frequency variation due to process, temperature, or non-linear capacitor bias point shift. Equation 2.3 shows the frequency sensitivity of an LC tank to capacitive variation.

$$\frac{\delta f}{\delta C_T} = \frac{-C^{-3/2}}{4\pi\sqrt{L}} \qquad (2.3)$$

For a tank defined by a 5nH inductance at 2GHz, a typical value for a fully-integrated oscillator, the frequency sensitivity is -856kHz/fF, nearly two orders of magnitude higher than the BAW resonator. Thus, an LC resonance would always need to be frequency locked to a reference even if perfect frequency accuracy were possible. In contrast, a BAW resonator, if sufficient accuracy were available, would not need to be frequency locked. This is a fundamental benefit of using high Q MEMS resonators in low power transceivers.

---

[1] And thus the subsequent oscillation frequency.

2.3 Circuit Proof-of-Concept I: 300μW Pierce Oscillator

To demonstrate the concepts presented in Sections 2.2 and 2.1, a proof-of-concept circuit was designed, implemented, and measured in a 0.18μm standard CMOS process [14] [19]. The goals of this project were threefold:

- A proof-of-concept circuit would verify the resonator/CMOS models and the co-design methodology. This verification is necessary as the accuracy of the models for MEMS components and weak inversion CMOS at RF frequencies are not as heavily developed as the traditional models for strong inversion CMOS circuitry, which designers have become accustomed to.
- A prototype would allow the refinement of the unresolved resonator/CMOS packaging and interconnect problem.
- This circuit, when used as a local oscillator, would also provide a stepping-stone to the implementation of an entire low power transceiver.

This section describes the state-of-the-art in transceiver local oscillator design. Used to generate the transmitted carrier frequency and the local oscillator (LO) signal, a stable, low-noise RF sinusoid generator is crucial for the performance of an RF link. Traditionally, this signal is obtained through frequency synthesis, which entails multiplying the frequency of a stable low frequency crystal oscillator via a phase- or delay- locked loop (PLL or DLL) [20]. There are, however, serious drawbacks with a frequency synthesizer. First, due to the low Q of the voltage controlled oscillator (VCO) tank and finite loop bandwidth of the PLL, the phase noise of the crystal oscillator is severely degraded by the frequency synthesizer. Secondly, the synthesizer consumes large amounts of power in the VCO and frequency dividers.

A few examples are useful for putting the problem in perspective. In a recent ultra low power frequency synthesizer design, approximately 400μW was consumed to provide a 434MHz carrier [7]. As the carrier frequency of these systems is increased into the GHz range, the power consumption of the frequency synthesizer increases dramatically. Outstanding phase noise performance can be achieved at the expense of high power dissipation. A recent high-performance 900MHz frequency synthesizer consumed 130mW with a phase noise of -127$\frac{dBc}{Hz}$ at a 330kHz offset [20]. The start-up time of a traditional frequency synthesizer is relatively long, and can be very inefficient if the transceiver requires agile duty-cycling and short packet transmission. Additionally, even with a "fully integrated" synthesizer, an off-chip quartz crystal is always required. Crystal oscillators typically exhibit very low phase noise due to the high quality factor of the crystal resonator. However, the resonant frequencies of quartz crystals are lower than most desired carrier frequencies, so frequency synthesis is usually required. A recent crystal oscillator implementation reports a phase noise of -113$\frac{dBc}{Hz}$ at 300Hz offset for a 78MHz oscillation frequency with a power dissipation of 340μW [21].

Another option is to use an integrated free-running LC-oscillator without a frequency reference. This could satisfy the need for a fast start-up time and low power dissipation, but the frequency variation and phase noise of such a design would be prohibitively poor. In [14], we presented an alternate method of sinusoid generation. A frequency reference is generated directly at the RF frequency of interest, with no low frequency reference. This is accomplished by placing a BAW resonator in the feedback path of a CMOS oscillator, ultimately combining the frequency stability of a mechanical resonance with the low power capability of standard submicron CMOS. The technique of co-designing the resonator with the CMOS electronics provides an extremely low power solution.

In this design, a single resonator is used to maximize the loaded Q of the oscillator. The impedance of the resonator is less than 5Ω at series resonance and larger than 1500 ohms at parallel resonance. Thus, to avoid severely loading the natural Q of the resonator, operation at the parallel resonance of the FBAR was chosen. The resonator occupies an area of approximately 100μm x 100μm and is wire-bonded directly to the CMOS chip containing the circuitry. The Pierce oscillator topology was chosen for its low phase noise potential and because it operates on the parallel resonance of the FBAR, allowing a higher loaded Q. A circuit schematic of the oscillator is shown in Figure 2.3. The signal is DC coupled to the first stage of the output buffer (M_{buf1}). Capacitors C_1 and C_2 represent the device, interconnect, and pad

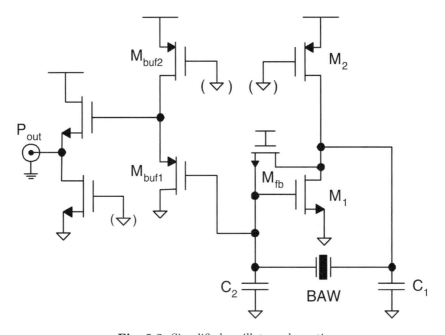

Fig. 2.3. Simplified oscillator schematic

capacitances. Accurate modeling and prediction of these values is crucial for the high frequency implementation of this topology. Transistor M_{fb} acts as a large resistor to provide bias to M_1. At the parallel resonance of the resonator, C_1, C_2, and C_o are tuned out. M_1 sees a high impedance at its drain node, allowing oscillation at this frequency. The sizing and layout of the circuit and resonator was optimized to minimize the power consumption of the oscillator.

At resonance, the initial loop gain is $A_L = g_{m1} R_L \frac{C_1}{C_2}$ where R_L is the real impedance seen at the drain of M_1 at the parallel resonance of the resonator. It can be shown that the optimal frequency stability and start-up factor are achieved with $C_1 = C_2$ [22]. Therefore, to minimize the transconductance necessary for oscillation and to maximize the output voltage swing for a given bias current, R_L was maximized. For a given frequency, BAW resonators may be designed with various membrane areas. As the area increases, the motional resistance (R_x) decreases. However, increasing the area also increases the feedthrough capacitance C_o. Increasing R_x degrades the loaded Q of the oscillator, thus decreasing R_L. Increasing C_o has a similarly detrimental effect. Thus, it is possible to calculate an optimal resonator area that minimizes the power dissipation and phase noise of the oscillator. Figure 2.4 plots R_L vs. the normalized resonator area for various values of $C_1 = C_2$. Using this technique, an optimal resonator area of approximately $(100 \times 100) \mu m^2$ was chosen for fabrication. The curve marker in Figure 2.4 indicates the design point. It is important to note that the finite Q of the CMOS device and pad capacitance must be taken into account, as they also reduce the loaded Q of the oscillator. This optimization led to a maximized value of R_L at parallel resonance. The desired voltage swing of the oscillator was 100mV zero-peak. The equation $V_0 = I_1 R_L$ relates the desired voltage swing to the first harmonic component of the drain current of M_1. Thus, since the oscillator is operated in the current-limited regime, the necessary oscillator core bias current is 300μA. As discussed in Section 2.1, weak inversion operation provides higher transconductance efficiency ($\frac{g_m}{I_d}$). The sizing of transistor M_1 $(500/0.18)\mu m$ was chosen as to provide subthreshold operation, ensuring sufficient initial transconductance for reliable start-up. For transistor M_1, $IC = 0.2$, yielding $\frac{g_m}{I_d} = 23$.

A symmetric resonator layout allowed equal loading on the drive and sense electrode. The resonator was wirebonded directly to the CMOS chip to eliminate board parasitics, which would drastically degrade the resonator response. To accomplish this, the CMOS and BAW pad layouts were constructed with equal spacing so the chips could be mounted in close proximity to each other and directly wirebonded. See Figure 2.5 for a photograph of the completed prototype. The two chips were bonded with conductive epoxy to a grounded substrate, resulting in A 200μm spacing between the chips. A custom assembly with two bondwires per interconnect was used in the initial prototype to reduce the bondwire inductance, but subsequent experimentation showed that standard chip-on-board (COB) assembly using one bondwire per pad and a 500μm chip spacing was adequate. The oscillator core was biased at

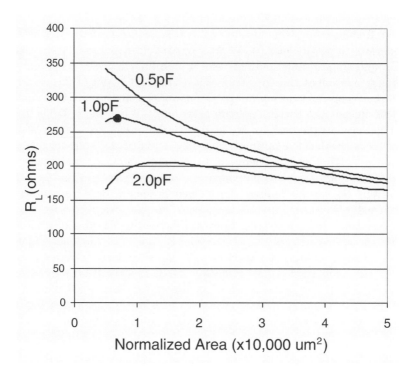

Fig. 2.4. Optimization of BAW resonator area. Three curves are shown with various values of $C_1 = C_2$

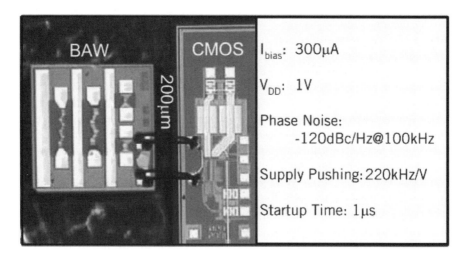

Fig. 2.5. Photograph of the CMOS/BAW prototype oscillator

300μA with a power supply voltage of 1V. The amplitudes of the 2nd, 3rd, 4th, and 5th harmonics were measured to be 30dB, 36dB, 45dB, and 49dB below the carrier, respectively. The measured phase noise performance is shown in Figure 2.6.

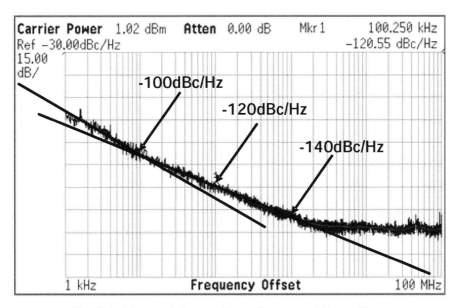

Fig. 2.6. Measured phase noise performance of the oscillator

Phase-noise was measured with an *Agilent E4445A PSA* and verified with an *HP 3048A* phase noise measurement system. The measured phase noise of the oscillator is -100 $\frac{dBc}{Hz}$ at 10kHz offset, -120 $\frac{dBc}{Hz}$ at 100kHz offset, and -140 $\frac{dBc}{Hz}$ at 1MHz offset. The start-up time of the oscillator was measured to be approximately 800ns, making it suitable as a frequency reference for receivers requiring agile duty-cycling. For example, it is possible to cycle the oscillator on/off between transmitted bits for low data rate OOK transceivers. Indeed, this property forms the basis for the transmitter that will be presented in Chapter 3. In addition, it was experimentally verified that analog pulse shaping of the OOK signal is possible by modulating the bias current of the oscillator as a function of time. Second, third, and fourth-generation oscillators have been fabricated and tested in a 0.13μm CMOS process, and similar performance results were observed. The supply pushing of the oscillator was measured to be 220$\frac{kHz}{V}$, yielding only a 26.4kHz frequency shift over a 10% supply voltage variation. The temperature coefficient of oscillation was also measured. A plot and detailed discussion of the effect of temperature on the oscillator is presented in Section 2.4, Figure 2.14. A useful figure-of-merit FOM for comparing the performance of RF oscillators is given by Equation 2.4 [23].

$$FOM = 10log\left[\left(\frac{\omega_0}{\Delta\omega}\right)^2 \frac{1}{L(\Delta\omega) \cdot V_{DD} \cdot I_{DD}}\right] \qquad (2.4)$$

This FOM was used to compare some recently published low power RF oscillators. Table 2.1 shows a breakdown of oscillator performance and FOM for these oscillators. This table depicts low power oscillators using a variety

Table 2.1. Comparison of recently published RF oscillators

Parameter	This work [14]	[7]	[23]	[24]
Power Consumption (μW)	300	230	1460	100
V_{dd} (V)	1	1	0.35	0.5
Oscillation Frequency (GHz)	1.9	0.4	1.4	1.9
Phase Noise ($\frac{dBc}{Hz}$ @ 1MHz)	-140	-118	-129	-114
FOM (dB)	210	177	190	190
Comment	BAW	SMT L	Integ. L	Bondwire L

of tuning elements, including surface mount inductors, on-chip integrated inductors, and bondwire inductors. As shown in the table, the co-design of BAW resonators and subthreshold-biased CMOS circuitry provides FOM performance at least two orders of magnitude above the state-of-the-art in RF oscillators.

Co-design and optimization of CMOS circuitry and RF MEMS components will become an increasingly important tool with the proliferation of this technology. The imminent integration of RF MEMS on the same silicon substrate as the CMOS will enhance the performance of the system as well as the need for co-design.

2.4 Circuit Proof-of-Concept II: Differential $300\mu W$ BAW-Based Oscillator

In Section 2.3, the co-design of a single-ended CMOS oscillator and BAW was presented. However, to achieve better supply rejection and higher output swings, a differential topology is often used. The goal of the work presented in this chapter was to design a differential BAW-based oscillator for direct comparison with the single-ended oscillator presented in Section 2.3. This section presents the design, implementation, and testing of a 300μW differential BAW based oscillator.

2.4.1 Analysis/Design

The goal of the differential oscillator is to drive the BAW resonator symmetrically and excite its parallel resonance while maintaining symmetry throughout the oscillator. In many traditional integrated differential oscillators, the negative resistance of the sustaining amplifier is achieved through a cross-coupled transistor pair. Looking into the cross-coupled pair, a wideband, DC-to-RF negative resistance is created. To achieve oscillation at a certain frequency, a parallel LC load is used to provide a high impedance at that frequency. At this frequency, the total tank impedance is negative and oscillation occurs. Bias current can conveniently be provided through the same inductors that tune out the tank capacitance.

A mechanical resonator-based differential oscillator, however, is somewhat more complicated. The following issues had to be overcome for successful implementation:

- **Low Frequency Stability:** The resonator can be shunted across the cross-coupled pair, providing a high Q response at the parallel resonance to set the oscillation frequency. However, at low frequencies, the resonator presents a high impedance. Thus, the circuit would be DC unstable and latch-up like a comparator.
- **Loaded Q:** Bias current must be supplied to the cross-coupled pair without de-tuning the BAW resonator at RF frequencies. Thus, a high impedance circuit environment at high frequencies is necessary.
- **Common-Mode Control:** A high-impedance bias circuit requires additional feedback circuitry to stabilize the common-mode of the oscillator. In traditional RF oscillator designs, the inductor tuning elements make common-mode feedback unnecessary.

Current sources could be used to supply bias current to the cross-coupled pair without de-tuning the resonator, but latch-up would occur. One way to circumvent this problem is to design a high-pass response into the cross-coupled pair negative resistance. This is realized by using separate current sources for the cross-coupled pair and coupling the sources through a capacitor [25]. At low frequencies, the cross-coupled pair experiences a large degeneration, reducing the negative resistance. At high frequencies, the sources interact, providing full transconducance from the cross-coupled pair. The simplified schematic of the oscillator is shown in Figure 2.7. It can be shown that the differential impedance looking down into the cross-coupled pair is given by Equation 2.5.

$$Z_{cc} = \frac{-1}{g_{m1}}[1 + \frac{g_{m1}}{s2C_s}] \tag{2.5}$$

Thus, at high frequencies, the structure provides $\frac{-1}{g_{m1}}\Omega$ of negative resistance. The effect of varying values of C_s is shown by the simulation results in Figures 2.8 and 2.9. Figure 2.8 shows the AC analysis of the oscillator loopgain for various values of C_s, swept logarithmically from 500fF to 5pF. The high Q BAW resonator series and parallel resonance is clearly visible. The desired oscillation mode is at the 1.9GHz parallel resonance peak. There is also a low Q low frequency resonance visible in the response due to the inductive nature of the capacitively degenerated cross-coupled pair interacting with the

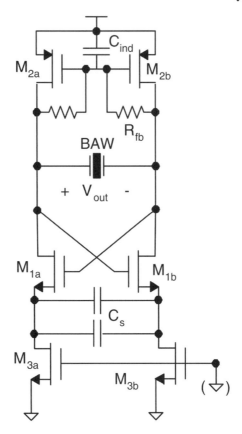

Fig. 2.7. Schematic of differential oscillator

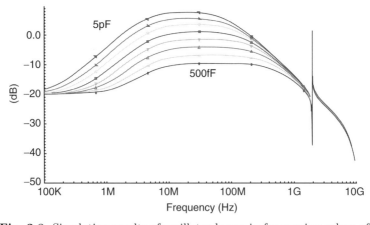

Fig. 2.8. Simulation results of oscillator loopgain for varying values of C_s

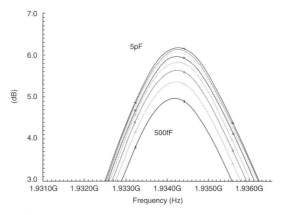

Fig. 2.9. Loopgain detail at oscillation frequency

BAW parallel plate capacitance. This resonance would cause a parasitic oscillation if the loopgain exceeded 0dB, and the stability would degrade as the parallel plate capacitance increased. To improve stability, C_s should be made small. However, reducing C_s increases the negative resistance pole frequency in Equation 2.5, thus reducing the oscillator loop gain at the desired resonance. Figure 2.9 shows a close-up of the oscillator loop gain at resonance for various values of C_s. The loop-gain degradation is clearly shown as C_s is decreased. Thus, proper choice of C_s is crucial for stable and efficient oscillation. For this oscillator, $C_s{=}1pF$ was chosen to provide stable operation over a range of BAW resonators.

The next difficulty is achieving a stable common-mode voltage equilibrium. The common-mode feedback must not degrade the high frequency differential impedance of the current sources. To achieve a low LF common-mode impedance and a high HF differential-mode impedance, a self-biasing common mode feedback circuit was developed. This structure consists of M_{2a}, M_{2b}, R_{fb}, and C_{ind}. At low frequencies, the common-mode impedance looking into the inductor is $\frac{1}{g_{m2}}$. At RF, the differential impedance is $R_{fb}//R_{o2}$. Thus, the structure provides a stable DC common-mode operating point but does not de-tune the high Q BAW resonator. One drawback of the structure is that it consumes substantial supply headroom, limiting the low voltage operation potential of the oscillator. The common-mode voltage is easily adjustable by varying the aspect ratio of transistors M_{2a} and M_{2b}.

The sizing of the cross-coupled pair is an important consideration for achieving low power consumption. For oscillation, the transconductance of each device is dictated by the following: $g_m > \frac{2}{R_{p,res}}$. The sizing is a tradeoff between transconductance efficiency and f_T degradation in the weak inversion regime. The devices were sized at $\frac{100\mu m}{0.13\mu m}$, yielding an inversion coefficient of 0.33, providing a $\frac{g_m}{I_d}$ of approximately 21.

2.4.2 Experimental Results (1.9GHz)

To verify these concepts, the oscillator was implemented in a standard 0.13μm CMOS process. The layout detail is shown in Figure 2.10. The oscillator layout

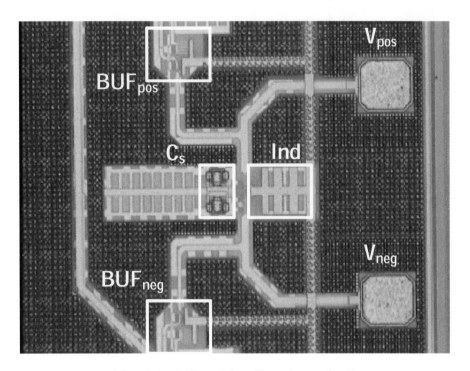

Fig. 2.10. Differential oscillator layout detail

was symmetric with a 290μm pad-to-pad spacing of the BAW interconnect pads. Both oscillator outputs are buffered with self-biased 50Ω output drivers that present a 100fF capacitance to the oscillator core. The active inductor structure and output buffers are visible in the photo. Capacitor C_s, also visible, was split into two cross-coupled capacitors to eliminate the asymmetry of the back plate connection. The oscillator layout is symmetric up to the buffer outputs, where an asymmetric pad structure was used for die area efficiency. A photograph of the oscillator COB assembly is shown in Figure 2.11. The board was assembled using standard COB techniques with no special processes required. The oscillator was then tested over various bias current levels. Oscillation was sustained at a bias current as low as 155μA total, corresponding to a g_m of approximately 1.6mS for each cross-coupled device. Thus, back-calculating the resonator parallel resonant impedance reveals $R_{p,resonator} = \frac{2}{g_m} = 1.2k\Omega$, agreeing well with calculations. The differential zero-peak swing for a 155μA bias current was 50mV. The output swing

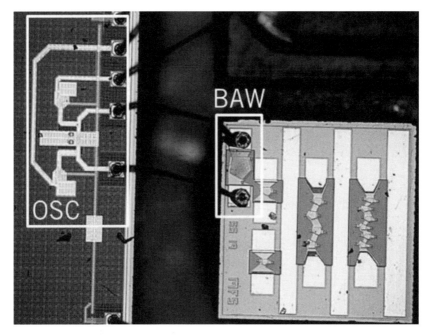

Fig. 2.11. Differential oscillator COB assembly

increased linearly with bias current up to a total bias of $600\mu A$, where the differential voltage swing was 318mV zero-peak. As the current was further increased, the oscillator entered the voltage-limited regime and subsequent increases in bias currents resulted in diminishing increases in voltage swing.

Next, the oscillator was biased to the design value of $300\mu A$ ($150\mu A$ through each leg) for the remainder of the testing, yielding a differential output power of -13.2dBm (corresponding to an oscillator core differential swing of approximately 200mV zero-peak). The phase noise was measured with an *Agilent E4445A PSA*. The data is plotted in Figure 2.12. A comparison of this phase noise measurement to that from the Pierce oscillator presented in Section 2.3 reveals interesting results. Since both oscillators operate with nearly equal differential voltage swings across the resonator (200mV zero-peak) and nearly equal resonator impedances, it follows that the power in the resonant tank is nearly identical. In addition, because the resonator quality factors are nearly equal, the phase noise of the both oscillators yield similar results when calculated with Leeson's formula.

Table 2.2 shows the phase noise comparison of the differential oscillator to the Pierce oscillator presented in Section 2.3.

Notice that the phase noise performances of the two oscillators are very similar. An oscillator supply-pushing performance of $500\frac{kHz}{V}$ was measured over a V_{dd} range of 0.7V to 1.4V. The low supply-pushing figure is due to the

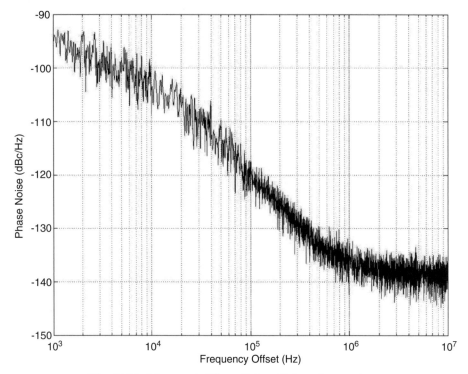

Fig. 2.12. Measured differential oscillator phase noise

Table 2.2. Measured oscillator phase noise $\left(\frac{dBc}{Hz}\right)$

f_{offset}(Hz)	Differential	Pierce
10k	-103.3	-100
100k	-120.3	-120
1M	-136.4	-140

low resonator frequency dependence to non-linear capacitor bias point shifts, as explained in Section 2.2. A 10% variation in the nominal 1.2V supply would result in only a 60kHz frequency shift.

The measured harmonic components of the differential and Pierce oscillators are compared in Table 2.3. As expected, the even order harmonics of the differential oscillator are suppressed by the differential architecture. The fourth harmonic was below the noise floor of the spectrum analyzer.

The start-up time of the oscillator was measured with an *Agilent 54855A* 8bit, 6GHz oscilloscope. Figure 2.13 shows the measured oscillator start-up transient. The positive and negative output voltages of the differential oscillator are shown. The start-up time, measured from the onset of bias current

Table 2.3. Measured oscillator harmonic distortion (dBc)

Harmonic	Differential	Pierce
Second	-55	-30
Third	-58	-36
Fourth	–	-45
Fifth	-65	-49

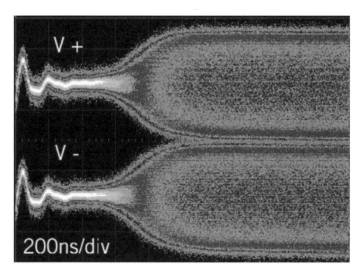

Fig. 2.13. Measured differential oscillator start-up transient

to oscillator saturation, is approximately $1\mu s$. This matches, both theoretically and experimentally, the start-up time of the Pierce oscillator presented in Section 2.3. This result is expected because both oscillators have similar tank quality factors and initial loop gains.

The temperature dependence of a BAW-tuned reference oscillator is important because it is not locked to a stable temperature compensated crystal reference. As described in [26], the temperature coefficient of the parallel resonance of a Molybdenum/Aluminum Nitride/Molybdenum (Mo/AlN/Mo) BAW resonator is approximately $-25\frac{ppm}{^\circ C}$. For a high Q resonant structure, the resonator temperature coefficient should determine the temperature coefficient of oscillation frequency (TCF). To verify this relationship, the differential oscillator frequency was measured over a temperature range of -20°C to 100°C. This measurement was performed with a *Thermonics T-2420* temperature forcing system. In addition, the Pierce oscillator that will be used in Chapter 3 was measured over the same range for comparison. See Figure 2.14 for the measured results. The differential oscillator uses a custom-designed 70Ω resonator while the Pierce uses a 50Ω resonator test structure. The theoretical

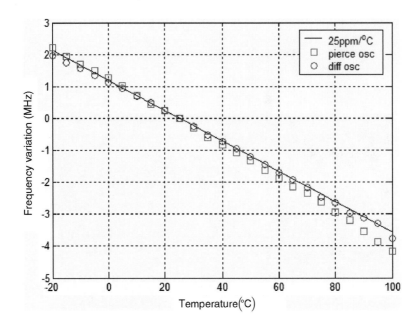

Fig. 2.14. Measured temperature coefficient of oscillation for two CMOS oscillators

$-25\frac{ppm}{^\circ C}$ curve was plotted for reference. As expected, the measured TCF of both oscillators closely match the theoretical resonator coefficient. The Pierce oscillator shows a slightly higher TCF that increases with temperature. This increase is due to the temperature dependence of the CMOS power amplifier that loads the oscillator. The addition of temperature-independent bias stabilization circuitry would return the curve to its theoretical TCF value. The linear, known TCF response may be easily compensated to provide a stable frequency reference. A compensation circuit providing approximately $-5\frac{fF}{^\circ C}$ to the resonator would provide temperature stability. In completely uncompensated form, both oscillators exhibit a +/- 1MHz tolerance from 5°C to 45°C.

The motional resistance of the BAW resonator also exhibits a temperature dependence [26]. Important to low power CMOS oscillators is the temperature coefficient of the parallel resistance, which affects the oscillator loop-gain, output swing, and phase noise performance. This temperature coefficient is approximately $-1500\frac{ppm}{^\circ C}$. Over a range of -20°C to 100°C, R_p decreases by approximately 18%. This relatively small variation is easily compensated by oscillator amplitude control circuitry.

2.4.3 Experimental Results (2.4GHz)

As designed, the differential oscillator can operate successfully over a wide frequency range with no modifications. One particularly interesting frequency band is the 2.4GHz ISM[2] band. This band is sufficiently high to allow completely integrated inductors, but low enough to allow subthreshold RF transistor biasing and low power consumption.

It should be noted that there is an optimal C_s for each oscillation frequency. As the frequency increases, the allowable cross-coupled pair pole frequency (given by Equation 2.5) can be increased, allowing a smaller C_s and thus higher stability against low frequency parasitic oscillations. A custom 2.4GHz resonator was designed and fabricated to demonstrate oscillator operation in this frequency band. Figure 2.15 shows a photograph of the completed ISM band oscillator implementation.

Fig. 2.15. ISM implementation of differential oscillator

The CMOS chip is shown below the BAW chip. The resonator chip was designed with five resonators. The four perimeter resonators are spaced in frequency across the ISM band. The resonator die (1mm x 1mm) was assembled on top of the CMOS die with non-conductive epoxy. Wirebonding provided

---

[2] Industrial, Scientific, and Medical

the electrical connectivity between the two chips. This experimental assembly methodology worked well and successfully de-coupled the CMOS/FBAR assembly and the CMOS/board assembly.

The oscillator was biased to $200\mu W$ per leg for stable startup ($400\mu W$ total). This relatively high bias current is due to the quadratic relationship of bias current to the operation frequency, which predicts a 1.6x increase in startup transconductance. In addition, due to the experimental nature of the ISM-band resonators, high variation in resonator motional resistances were expected. Figure 2.16 shows the steady-state transient oscillator output. Oscillation occurred at 2.43GHz as expected. The oscillator exhibited a clean

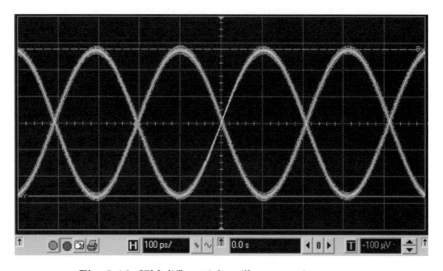

Fig. 2.16. ISM differential oscillator transient output

spectrum and good amplitude matching. The phase-noise of the oscillator was measured with an *Agilent E5052A* Signal Source Analyzer over four bias point settings ($400\mu A$, $500\mu A$, $600\mu A$, $700\mu A$). See Figure 2.17. As expected, the close-in phase noise and noise floor decrease as the oscillator loop power increases. At a bias current of $400\mu A$, the oscillator exhibits a phase noise of approximately $-113\frac{dBc}{Hz}$ at 100kHz offset. This is approximately 7dB higher than the previous 1.9GHz version of the oscillator, indicating a reduced resonator quality factor. The inferred quality factor of the resonator from these phase noise measurements is approximately 200.

This proof-of-concept demonstrates the practical application of MEMS and integrated circuit co-design techniques at higher frequency bands. Our current work in this area involves the design of a quadrature voltage controlled oscillator (QVCO) tuned by BAW resonators. This work will involve coupled RF oscillators to generate accurate quadrature sinusoids. Although the phase

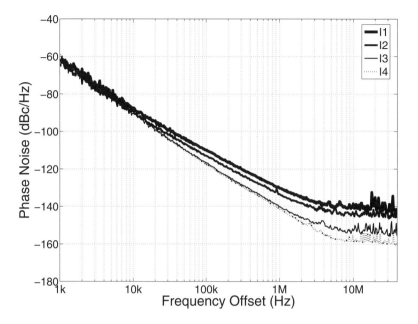

Fig. 2.17. ISM differential oscillator phase noise

noise performance will be significantly better than a traditional LC-tuned oscillator, a number of difficulties remain. First, the tuning range of high Q structures is fundamentally very low. To allow a reasonable tuning range, it is anticipated that a switched resonator array will be necessary in addition to capacitive tuning. Secondly, maintaining low phase error between the two oscillator cores is crucial. This phase error is determined (in part) by the center frequency offset the two resonators. When a high-Q resonant structure is used, the frequency alignment specifications become stringent, so adaptive calibration techniques are currently being explored to maintain high phase accuracy.

2.5 System Proof-of-Concept: Energy Scavenging Transmit Beacon

The previous section confirmed the design philosophies involved in co-designing BAW resonators and CMOS circuitry at RF frequencies. This section seeks similar confirmation of the system-level philosophies: *Can 1cm³ energy scavengers realistically and robustly power GHz-range RF circuitry?* The ultimate goal of this work, as reported in [27] was to develop a completely self-powered wireless node.[3] The two most applicable energy scavenging technologies are solar power and vibration-based power, due to the large potential application space of these technologies. Photovoltaic solar cells are a mature technology, and a solar cell based power source may be implemented using commercial off-the-shelf technology. In addition, the technique of utilizing low-level vibrations as a power source was also investigated.

This section reports a 1.9GHz transmit beacon that is successfully operational using solar and vibrational power sources. The main components of the design are the energy scavenging devices, the powertrain, the local oscillator, the power amplifier, and the antenna. One main goal was to maximize the efficiency of the conversion from solar and vibrational power to transmitted RF power. The power circuitry converts the scavenged energy into a stable supply voltage for the RF circuitry. This process entails converting a high impedance, unstable supply into a stable, low impedance 1.2V supply. A $10\mu F$ storage capacitor functions as an energy reservoir. When the capacitor charges to a pre-specified energy level, the supply rails to the RF circuitry are activated and energy is consumed. Because the transmitter dissipates power faster than the rate at which the piezoelectric generator or solar cell can produce it[4], the voltage across the storage capacitor falls when the radio is on. Once the energy has been depleted to a level specified by a "Shutdown control" block, the supply rails are disabled and the capacitor is recharged.

The choice of energy storage mechanism involves a tradeoff between energy density and reliability. Batteries have far higher energy density than do capacitors. For example, rechargeable lithium ion batteries have an energy density of roughly 1000 J/cm^3. Ceramic capacitors have an energy density on the order of 1 to 10 J/cm^3. However, most lithium ion batteries are limited to 500 to 1000 recharge cycles and have a finite shelf life. Furthermore, the life of the battery would suffer in the proposed application because the batteries are kept charged with a trickle of current rather than undergoing deep discharge. Capacitors, on the other hand, have an almost infinite lifetime and are simpler to charge. Because wireless sensor nodes do not require substantial energy storage, and because lifetime of the node is a primary concern, a capacitor was used.

Although a linear regulator would result in a lower total efficiency in the current design scenario, a linear regulator was chosen because of its ability for a higher level of integration and increased simplicity[5]. The transmitter was designed using the same CMOS/MEMS co-design philosophies described earlier in this chapter. A photo of the transmitter is shown in Figure 2.18. The entire transmit area consumes less than 20mm^2 of board space. The oscillator

---

[3] The work in this section was in collaboration with S. Roundy, Y.H. Chee, and P. Wright

[4] Except in direct sunlight

[5] No external filtering inductors or capacitors is needed, in contrast to switching regulators

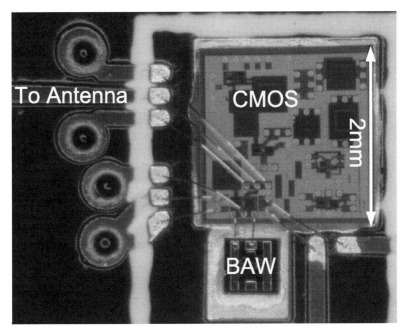

Fig. 2.18. RF Transmitter COB implementation

provides a 100mV signal to the integrated power amplifier, which provides a 0dBm signal to the chip antenna. An output power of -1.5dBm was measured in close proximity to the chip antenna. The completed $(2.4 \times 3.9)\text{cm}^2$ transmit beacon implementation is shown in Figure 2.19. The powertrain, photovoltaic solar cell, transmitter, and 1.9GHz chip antenna are clearly visible. All circuitry is placed on the back of the board, allowing the front to be utilized for solar collection and RF emission. Although a commercial, off-the-shelf voltage regulator was used, a custom integrated circuit could provide higher efficiency and a much smaller board footprint[6]. A custom piezoelectric bender was also designed[7] to power the beacon. The bender exhibits a resonant frequency of approximately 200Hz and a quality factor of about 20. Thus, the output power drops off rapidly as the stimulus frequency drifts from the bender resonant frequency. Although not shown, the transmit beacon was successfully powered by this bender, demonstrating that hybrid energy scavenging is a viable option for wireless sensor networks.

See Figure 2.20 for the measured beacon voltage levels under low lighting conditions. V_{supply} is the voltage on the storage capacitor. Under low solar input conditions, this voltage slowly charges to a pre-defined level. At that voltage, the voltage regulator is enabled and V_{reg} rises to a stable 1.2V,

---

[6] This is a subject of current research
[7] Designed by Shad Roundy, U.C. Berkeley [3]

Fig. 2.19. Transmit beacon implementation

powering the RF transmitter. The RF blocks are then enabled, beginning transmission. During this stage, V_{supply} decreases as energy is drawn from the storage capacitor. When a sufficiently low level of capacitor energy is sensed, the voltage regulator is disabled and the RF transmission stops. The main figure of merit of this system is the transmit duty cycle under various environmental conditions.

The achievable transmit duty cycle varies dramatically over various lighting conditions. In the presence of low indoor light, the duty cycle is approximately 0.4% (400bps throughput assuming a 100kbps transmitter datarate). In direct sunlight, the duty cycle is 100% (in that condition, the solar cell is supplying more power than necessary for constant transmitter operation). For a $5.7m/s^2$ vibration level (about $0.58G$), the transmitter duty cycle is about 2.6%, corresponding to a throughput of 2.6kbps. The transmit beacon operates indefinitely from ambient energy and has been in continuous service for approximately 3 years.

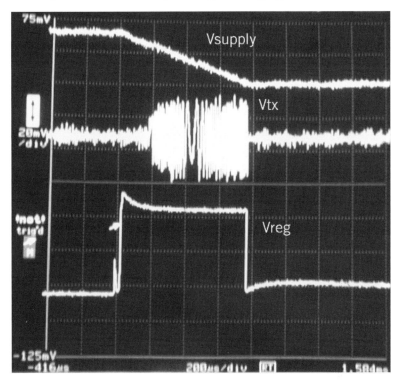

Fig. 2.20. RF transmit beacon under low light conditions

3

TWO CHANNEL BAW-BASED TRANSCEIVER

This Chapter describes the implementation of a low power, fully integrated, two-channel transceiver that builds off the principles that were developed in Chapter 2. For example, Chapter 2 discussed how RF MEMS technology and weak inversion standard CMOS may be used to reduce the power consumption and increase the integration of low power RF components. Two proof-of-concept designs were described: a low power oscillator and an entire energy scavenging transmit beacon. Building on this knowledge, an entire transceiver based on these concepts was designed. The following goals set the landscape for the decisions made during the design of the transceiver.

- Design a transceiver that leverages the benefits of state-of-the-art RF MEMS technology *and* incorporates a vision of future technology scaling of RF MEMS.
- The transceiver should be robust against BAW process variations.
- The effort should result in a working ultra-low power transceiver that can be used in peer-to-peer wireless sensor networks and represents a reduction in power consumption over the current state-of-the-art.

This chapter describes how these goals were met as well as lessons learned during the course of the design. This transceiver was presented in [28].

3.1 Architecture

The receiver block diagram is shown in Figure 3.1.

The two-channel tuned radio frequency (TRF) architecture was chosen to demonstrate the effectiveness of RF MEMS resonators in low power transceivers. The antenna feeds a 50Ω impedance presented by the LNA. The LNA drives a tuned LC load absorbing the capacitive input of two channel select amplifiers (CSAs). Each CSA incorporates an FBAR resonator, which performs receiver channel selection. Although two channels were used in this

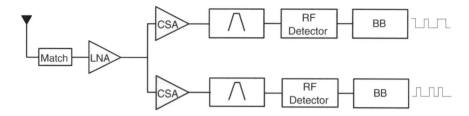

Fig. 3.1. Block diagram of the two-channel transceiver

implementation, the architecture is scalable to larger numbers of channels. In the future, advanced RF MEMS technology will allow lithographically defined resonant frequencies and the integration of MEMS onto CMOS wafers. At that point, this architecture could accommodate multiple channels with widely varying frequencies, limited only by the bandwidth of the LNA[1]. The CSAs drive an envelope detector, which acts as a self-mixer to perform signal downconversion. Baseband buffers are included to drive test instrumentation. Because the frequency stabilization is performed entirely by the MEMS resonators, no quartz crystals were used in this receiver architecture. The absence of a phase-locked loop (PLL) ensures a much faster receiver start-up time than that required by a traditional radio.

The two-channel embodiment displays flexibility in terms of modulation schemes: the receiver can detect two unique on-off keyed (OOK) data streams at two carrier frequencies, or it can detect a frequency shift keyed (FSK) modulation. For dense wireless sensor networks, it is anticipated that two separate OOK channels will be used, with one reserved for beaconing. Changing between these two modulation schemes can be accomplished with no receiver modifications, and can be performed dynamically in either the analog or digital baseband detection circuitry.

3.2 LNA Design

The low noise amplifier (LNA)[2] was designed to serve three main purposes:

- Present a 50Ω impedance match to the antenna.
- Absorb the capacitive load of multiple (N) channel select amplifiers. For a scalable architecture, it is important that the addition of each subsequent channel results in little (or no) performance degradation of the amplifier.
- Achieve high RF gain to overcome the large noise of the self-mixing process.

---

[1] Approximately 200MHz
[2] Designed by Richard Lu, EECS Dept., U.C. Berkeley

A single-ended, cascoded, tuned input/output LNA was chosen to meet these requirements with a minimum amount of power consumption. See Figure 3.2 for a front-end schematic of the receiver.

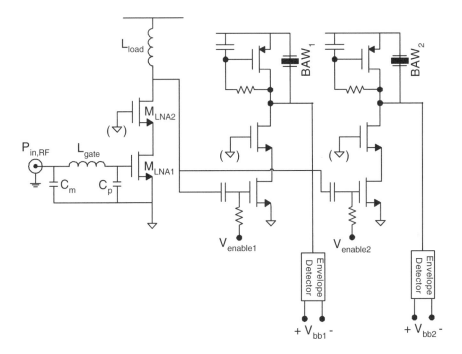

Fig. 3.2. Simplified receiver front-end schematic

The LNA comprises transistors M_{LNA1} and M_{LNA2}. Moderate inversion operation of M_{LNA1} allowed a reasonable inductor value of L_{gate}, providing the possibility of implementing this inductor on-chip. As the f_T of CMOS processes continues to scale, the size of this inductor will increase to prohibitively large values. Operation in weak inversion is a useful way to control this trend. Although the f_T is degraded, the performance penalty is partially mitigated by the increase in transconductance efficiency. The output tank consists of L_{load}, the channel select amplifier capacitance, and a small explicit capacitor to achieve the correct resonant frequency.

A standalone LNA test structure was implemented for performance verification. The output was matched through a passive capacitance transformer for testing purposes. Figure 3.3 shows the measured input matching and forward gain of the amplifier when biased at 1.4mA.

The measurement shows good input matching and correct output tank frequency. The S_{21} of the standalone LNA at resonance is approximately 13dB,

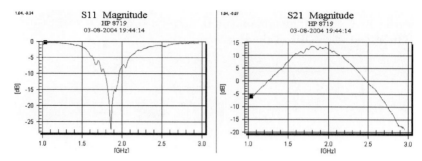

Fig. 3.3. Measured S-parameters of the LNA test structure

corresponding to an in-situ[3] voltage gain of 22dB. The amplifier noise figure, measured with an *HP 8970B* Noise Figure Meter, is 2.5dB at 1.9GHz. See [29] for a thorough discussion of the design methodology and performance of the LNA.

3.3 CSA Analysis and Design

The channel select amplifier performs two main functions: it provides high RF gain in order to overcome the high noise figure of the detector. In addition, it interfaces the electrical signal with the acoustic resonator to perform high Q filtering of the signal. The amplifier must exhibit high RF gain with low power consumption while limiting the extent to which the resonator is de-tuned. Three separate RF amplifiers were implemented on each chip: one standalone test structure (50Ω input/output) and two integrated amplifiers; one for each of the two radio channels. This section discusses the design of the active choke inductor, the standalone amplifier, and the integrated amplifier.

3.3.1 Active Inductor

The BAW resonator, used in the parallel resonance mode, performs high Q filtering by presenting a high impedance for a very narrow bandwidth about its parallel resonant frequency. However, for off-resonance frequencies, the resonator presents a $\frac{1}{j\omega C}$ impedance[4]. Thus, at low frequencies (LF), the resonator presents a high impedance to the active circuitry. To provide bias current (a low DC impedance) to the active devices and reduce LF gain, a low load impedance at LF is desirable. However, at the signal frequency, the impedance must be high to avoid de-tuning the BAW resonator. Thus, an inductive structure is a natural candidate. To avoid substantial de-tuning of

---

[3] Meaning in the receiver, loaded with a high impedance LC tank

[4] Mechanically, off-resonance, the resonator is simply two parallel plates filled with an AlN insulator with relative permittivity $k = 9$

the BAW resonator at 1.9GHz, the bias device must present an impedance greater than 1500Ω. This specification corresponds to an inductance of 125nH, which is marginally reasonable to fabricate on-chip. Even if possible, it would resonate with the parallel plates of the BAW at (2pF) at 318MHz, adding an out-of-band response to the receiver input. It would also consume more than 10,000μm² of silicon area. The use of an active inductor, however, allows the realization of very high inductance values and easy control over the inductor Q. For this application, a large, low Q inductance is needed to provide high RF impedance with a low frequency, low Q parasitic resonance.

An active inductor topology is proposed for this application. See the inductor schematic in Figure 3.4. At low frequencies, the inductor impedance is $\frac{1}{g_{m,ind}}$. At high frequencies, the impedance is $R_{o,Mind}//R_{fb}$. It can be shown that the effective inductance and Q of the structure are given by Equations 3.1 and 3.2, respectively.

$$L = \frac{R_{fb}C_{ind}}{g_m} \tag{3.1}$$

$$Q = \omega R_{fb}C_{ind} \tag{3.2}$$

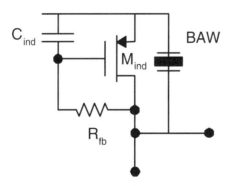

Fig. 3.4. Schematic of active choke inductor structure

The active inductor had two main design requirements. First, the noise contribution to the overall amplifier noise figure must be small. Second, the active inductor must be synthesized using reasonable passive component values that consume little silicon area. Table 3.1 shows the design values for the active inductor that will be used in subsequent calculations. As shown in the table, the active inductor achieves LF and HF impedances of approximately 360Ω and 6kΩ, respectively. The parasitic resonance of the active inductor and the resonator plate capacitance occurs at 34MHz with a Q of 6.4. Cascaded with the LNA presented in Section 3.2, the antenna-to-detector response in the 34MHz band will be negligible. This would not be the case with a 100nH on-chip choke inductor. Active inductors are routinely believed

Table 3.1. Active inductor design values

Component	Design Value
$R_{p,resonator}$	1.4kΩ
$I_{bias,CSA}$	500μA
$g_{m,ind}$	2.8mS
$\frac{g_m}{I_d}$	5.6
C_{ind}	5pF
R_{fb}	6kΩ
L	8.5μH
C_{res}	2pF
$f_{res,calc}$	38MHz
$f_{res,sim}$	34MHz
Q_{34MHz}	6.4

to have a prohibitively high noise figure. However, in this application, when used as a choke inductor, it will be shown that the excess noise that accompanies the structure is sufficiently low. The noise performance of this structure will now be analyzed. As shown in Equation 3.3, the noise is first calculated considering only the resonator[5].

$$\overline{v_0^2} = 4KTR_{p,resonator} = 22.4\frac{(nV)^2}{\sqrt{Hz}} \tag{3.3}$$

$$Simulation : 24\frac{(nV)^2}{\sqrt{Hz}}$$

As shown in Equation 3.4, the resonator is de-tuned by the accompanying circuitry. Additional components also modify the noise characteristics. Finally, equation 3.6 includes the transistor noise sources.

$$R_{p,effective} = R_{p,resonator}//R_o//R_{fb} = 1k\Omega \tag{3.4}$$

$$\overline{v_0^2} = 4KTR_{p,effective} = 16\frac{(nV)^2}{\sqrt{Hz}} \tag{3.5}$$

$$\overline{v_0^2} = \left[4KT\frac{1}{R_{p,effective}} + 4KT\gamma g_m\right]R_{p,effective}^2 = 45\frac{(nV)^2}{\sqrt{Hz}} \tag{3.6}$$

$$Simulation : 39.3\frac{(nV)^2}{\sqrt{Hz}}$$

The load noise factor, (β), is given by Equation 3.7.

---

[5] Corresponding to a fictitious circuit providing an ideal, noiseless current source and common-mode feedback

$$\beta = \frac{\left[4KT\frac{1}{R_{p,effective}} + 4KT\gamma g_m\right]R_{p,effective}^2}{4KTR_{p,effective}} = 1 + \gamma g_m R_{p,effective} \quad (3.7)$$

$$\beta = 2.86$$

Because the load contributes little noise to the total noise figure, the excess noise factor is acceptable for this application.

3.3.2 Standalone RF Amplifier

The RF amplifier consists of an NMOS cascode transconductance stage and a tuned load. A cascode transconductor structure is used to increase reverse isolation, ensuring amplifier stability. The tuned load consists of an FBAR resonator to perform channel selection and the inductive PMOS-R-C structure described in Section 3.3.1, which stabilizes the low-frequency bias point of the amplifier. The loaded quality factor of the tuned load is approximately 600, yielding an RF bandwidth of 3MHz.

A standalone RF amplifier was designed and implemented to verify the performance of this structure. Due to RF testing considerations, the structure had to be modified slightly from the in-situ receiver implementation. Figure 3.5 for the schematic of the standalone RF amplifier. Transistor M_1 is operated in weak inversion for two main reasons: first, the degradation in f_T is not crucial because the gate capacitance of the integrated version is tuned out by the LNA. Secondly, the voltage gain of the stage is proportional to the transconductance of M_1, and $\frac{g_m}{I_d}$ is maximized in lower levels of inversion. In this implementation, a $\frac{g_m}{I_d}$ of approximately 20 is achieved. The gate of cascode device M_2 is self-biased, and the layout of M_1 and M_2 is optimized for low drain and interconnect parasitic capacitance.

The input is terminated with an on-chip 50Ω resistor. This allows accurate RF characterization by minimizing the reflections coming from the 50Ω instrumentation transmission lines. An on-chip buffer is used to present a 50Ω output impedance, which is accomplished by maintaining the g_m of M_{2buf} at 20mS. The noise figure of the standalone amplifier can be calculated as follows:

$$F = \frac{SNR_{in}}{SNR_{out}} = \frac{S_{in}}{S_{out}}\frac{N_{out}}{N_{in}} = \frac{1}{A_v^2}\frac{\overline{v_o^2}}{4KTR_s} \quad (3.8)$$

Equations 3.9 and 3.10 show the voltage gain and total output noise, respectively.

$$A_v = \frac{R_{in}}{R_s + R_{in}}g_{m1}R_{p,effective} \quad (3.9)$$

$$\overline{v_o^2} = 4\left(\frac{4KTR_s}{2}\right)A_v^2 + (4KT\gamma g_m)R_{p,effective}^2 + \beta 4KTR_{p,effective} \quad (3.10)$$

Thus, the noise figure is given as Equation 3.11

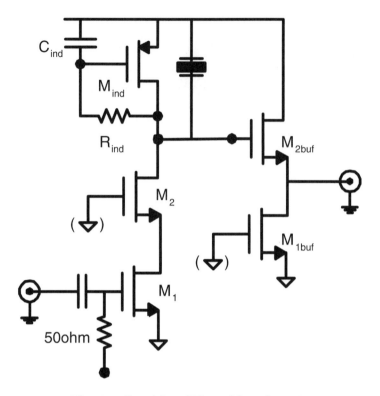

Fig. 3.5. Standalone RF amplifier schematic

$$F = \frac{1}{\left(\frac{R_{in}}{R_s + R_{in}} g_{m1} R_{p,effective}\right)^2} \cdot$$

$$\frac{4\left(\frac{4KTR_s}{2}\right) A_v^2 + (4KT\gamma g_m) R_{p,effective}^2 + \beta 4KT R_{p,effective}}{4KTR_s}$$

(3.11)

which simplifies to Equation 3.12.

$$F = 2 + \frac{4\gamma}{g_m R_s} + \frac{R_{p,effective}\beta}{R_s A_v^2}\Big|_{R_s = R_{in}}$$

(3.12)

$$F = 11.7dB$$

Table 3.2 shows the breakdown of the noise in the various amplifier components. A majority of the noise is due to the main transconductor. This is due to the relatively high noise figure of the standalone, 50Ω referenced test

Table 3.2. Breakdown of RF amplifier noise sources

Noise Source	Noise Contribution
Transistor M_1	60.8%
Source Resistance	21.0%
Active Resonator Load	18.2%

structure. Less than 20% of the noise contribution is due to the active choke inductor. The simulated S_{21} and NF are shown in Figure 3.6.

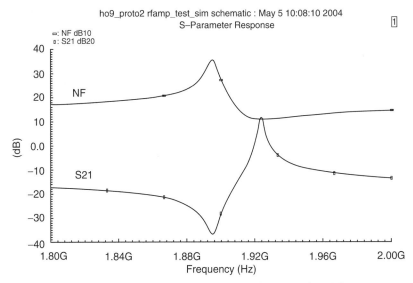

Fig. 3.6. Simulated S_{21} and NF of the RF Amplifier

The series and parallel resonances of the BAW resonator are clearly visible in Figure 3.6. At the parallel resonance the simulated noise figure[6] is 10.8dB and the simulated S_{21} is 11.5dB. The implemented standalone amplifier is shown in Figure 3.7.

Clearly visible are the three wirebonds connecting the CMOS chip to the resonant structure in a ground-signal-ground (GSG) configuration. The chips were placed in close proximity to allow a short bondwire length. This increases the frequency of parasitic resonances due to the bondwire inductance, eliminating the possibility of parasitic responses in-band. The amplifier core was biased at the design value of $500\mu A$. The measured S-parameters of the standalone amplifier are shown in Figure 3.8.

---

[6] Including the loss from the output buffer. This corresponds to an in-situ voltage gain of 17.5dB.

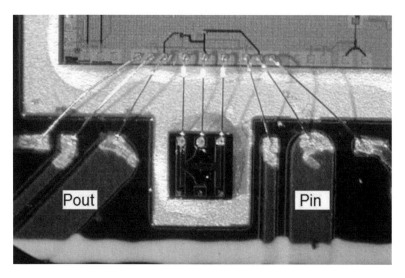

Fig. 3.7. Photograph of standalone RF amplifier

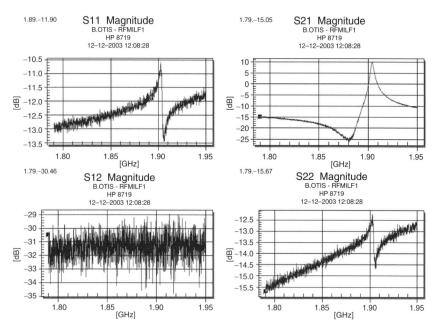

Fig. 3.8. Measured S-Parameters of the RF Amplifier

Because the standalone test structure is terminated by an on-chip resistor at the input and a on-chip buffer at the output, the S_{11} and S_{22} shows matching <-10dB. The S_{21} measurement shows good bandwidth agreement with the simulated and calculated values. The measured gain at parallel resonance is 10dB. The 2dB discrepancy can be attributed to variations in resonator R_p and board trace losses. The noise figure measurement was performed with an *HP 8970B* Noise Figure Meter and was duplicated using the "Spectrum Analyzer Method" with an *Agilent E4445A PSA*. A summary of the calculated, simulated, and measured parameters is shown in Table 3.3.

Table 3.3. Table of RF amplifier parameters

	NF	S_{21}
Calculated	11.7dB	17.7dB
Simulated	10.8dB	17.6dB
Measured	11.4dB	16.0dB

Close matching was observed between all calculated, simulated, and measured circuit parameters.

3.3.3 In-Situ RF Amplifier

The circuit design of the in-situ amplifier is the same as the standalone version except there is no need for 50Ω input/output matching. The input is AC coupled to the LNA, and the output is DC coupled to the RF detector. See Figure 3.9 for the schematic of the in-situ amplifier. Although the circuit topology is the same, the noise performance of the in-situ RF amplifier differs greatly from the standalone RF amplifier, due to the significantly higher source resistance of approximately 500Ω. This manifests itself as a much lower effective noise figure. See Equation 3.13 for the output noise power of the amplifier.

$$\overline{v_o^2} = 4KTR_sA_v^2 + (4KT\gamma g_m)R_{p,effective}^2 + 4KTR_{p,effective}\beta \qquad (3.13)$$

The noise figure is thus:

$$F = \frac{1}{(g_{m1}R_{p,effective})^2} \frac{4KTR_sA_v^2 + (4KT\gamma g_m)R_{p,effective}^2 + 4KTR_{p,effective}\beta}{4KTR_s} \qquad (3.14)$$

Which reduces to:

$$F = 1 + \frac{4\gamma}{g_m R_s} + \frac{R_{p,effective}\beta}{R_s A_v^2} \qquad (3.15)$$

Notice that the first term in Equation 3.14 is unity, not two as in Equation 3.12. This 6dB noise penalty is not present in the in-situ amplifier because

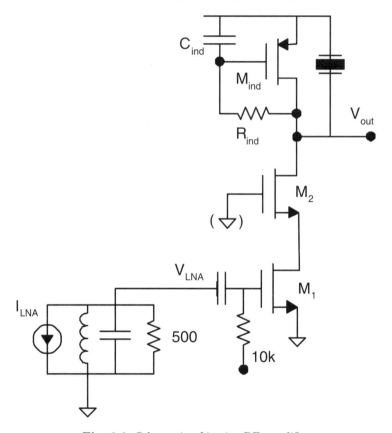

Fig. 3.9. Schematic of in-situ RF amplifier

the input is not resistively terminated. Additionally, because R_s is 10x higher than before, the effective noise figure of the amplifier is much lower.

$$F = 1.15 dB$$

Table 3.4 shows the noise breakdown for the two amplifier topologies.

Table 3.4. Breakdown of the standalone and in-situ amplifier noise sources

Noise Source	Standalone	In-Situ
Transistor M_1	60.8%	17.8%
Source Resistance	21.0%	76.7%
Active Resonator Load	18.2%	5.5%

The in-situ amplifier's main transconductor contributes less than 20% of the total output noise, compared to over 60% from the standalone amplifier. The active resonator load of the in-situ amplifier contributes negligible noise to the output. Thus, integrated into the receive chain, the RF amplifier contributes negligible noise to the receiver noise figure.

3.4 RF Detector Analysis and Design

The RF detector performs a signal level detection on each channel of the receiver. Rectification, or self-mixing, is achieved through the non-linear low-pass filtering of the RF input signal. The self-mixing is achieved through MOS-FETs biased in the deep-subthreshold region to increase their non-linearity. The simplified schematic of the detector is shown in Figure 3.10.

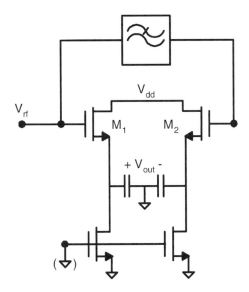

Fig. 3.10. Schematic of the RF Detector

This circuit, which was previously used in bipolar applications [30], is well suited to subthreshold CMOS implementations as well. Transistors M_1 and M_2 are sized at $\frac{10\mu m}{1\mu m}$ and biased at $\frac{10nA}{\mu m}$ (IC $= 0.015$), yielding a $\frac{g_m}{I_d}$ of approximately 30. The detector exhibits an intrinsic low-pass filter with a cutoff frequency set to 300kHz, which attenuates all fundamental tones passing through the receive chain. Although the receiver front-end is single-ended to reduce power consumption, it is desirable to have a fully-differential baseband for increased supply rejection and dynamic range. To perform a single-ended to pseudo-differential conversion, a replica envelope detector with feedforward offset cancellation was implemented. The bandwidth of the offset cancellation filter is 1.5MHz. The current consumption of each detector, including replica circuitry, is 200nA. Additionally, the efficiency of the circuit can be increased if transistor M_1 and its replica are designed in a DTMOS configuration (gate connected to body contact) [31]. This configuration decreases the subthreshold slope to a value approaching 60mV/decade, increasing the conversion gain of the detector.

A standalone envelope detector was designed and fabricated to verify its performance and conversion gain. Figure 3.11 shows the measured transfer function of multiple envelope detectors.

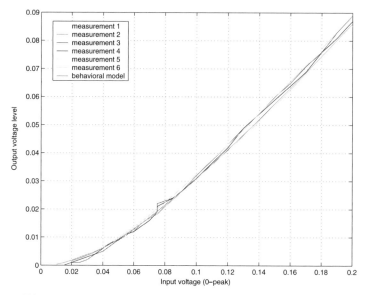

Fig. 3.11. Measured conversion gain of the envelope detector

These results show the decrease in conversion gain that occurs at low input signal levels. This creates a thresholding effect, where decreasing RF input levels cause a dramatic decrease in the noise figure of the detector. This threshold sets the noise floor of the tuned RF receiver.

Each channel in the receive chain also contains a low-power buffer to drive off-chip instrumentation[7]. The buffers are capable of driving a 20pF/2kΩ off-chip load while consuming 50μA. On-chip threshold-referenced bias circuits provide a moderate level of power-supply independence.

---

[7] Designed by Nathan M. Pletcher, EECS Dept., U.C. Berkeley

3.5 Transmitter Architecture and Design

In a typical sensor network, the transmitter sends out sporadic bursts of short data packets to neighboring sensor nodes (<10m). As shown in Figure 1.2, for a receiver sensitivity of -70dBm and a healthy margin for indoor multi-path fading conditions, a transmit power of approximately 0dBm is required. The transmitter must exhibit fast turn-on time and high efficiency. The transmitter architecture shown in Figure 3.12 is well-suited for these requirements.

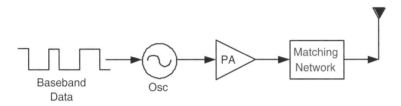

Fig. 3.12. Block diagram of the transmitter

The architecture utilized direct modulation of the BAW-based oscillator. The oscillator is similar to the one presented in Section 2.3. Migration from the 0.18μm to the 0.13μm technology node, however, required a re-optimization of the transistor sizing. The start-up time of the oscillator is approximately 1.5μs, allowing direct modulation of the transmitter to datarates over 150kbps. Direct modulation eliminates power hungry mixers and PLLs. Multiple channels can be implemented by tuning the oscillation frequency or by adding oscillators/transmit chains in parallel. For this implementation, two separate BAW resonators were used, so two discrete Tx chains were implemented. The power amplifier (PA) is a standard class-AB cascoded topology[8]. The high PA drain impedance is matched to the 50Ω antenna through a fully integrated, high Q capacitive transformer. See Figure 3.13 for a simplified transmitter schematic.

3.6 Experimental Results

3.6.1 Implementation

The transceiver was implemented in a standard 0.13μm CMOS process. The complete transceiver system is shown in Figure 3.14. The chip area is (4×4)mm^2, which is mostly consumed by passive test structures. The actual area consumed by the transceiver is approximately 8mm^2. Chip-on-board

---

[8] Designed by Yuen-Hui Chee, EECS Dept., U.C. Berkeley

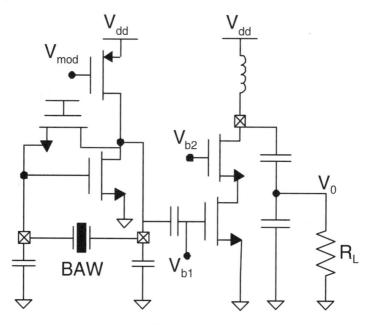

Fig. 3.13. OOK Transmitter Schematic

(COB) wirebonding was used to interface to the chip. As shown in the system photograph, the four FBAR resonators are bonded directly to the chip using standard COB wirebonding. This technique eliminates board parasitics on these sensitive nodes and reduces the required board area. The testboard was connected to a single 1.2V supply for testing.

3.6.2 Receiver

The normalized receiver gain of both channels is shown in Figure 3.15.

Both receive channels exhibit a 3MHz bandwidth and close gain matching. This implies that, when integrated into the receive chain, the BAW resonators in the channel select amplifiers are not significantly de-tuned. Receiver sensitivity for a 12dB SNR was measured at -78dBm[9]. The start-up time of the receiver is crucial in a wireless sensor network application due to the very short packet lengths. Figure 3.16 shows the receiver start-up transient. The bottom trace is the 0 to V_{dd} enable signal transient, while the top trace is the baseband output in the presence of a 20kbps OOK input signal. The measured start-up time is 10μs, minimizing the overhead associated with heavy duty-cycling. The total receiver current consumption is 3mA from a 1V supply with

---

[9] Receiver sensitivity was measured with external baseband amplification to overcome the noise of the test equipment

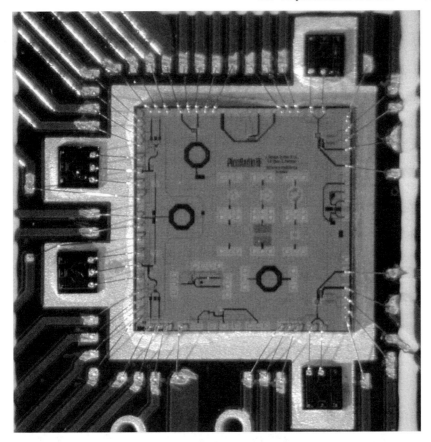

Fig. 3.14. Photograph of the transceiver implementation

both channels active. See Figure 3.17 for a component current breakdown of the receiver. Clearly, a majority of the current is consumed by the LNA and channel select amplifiers. With only one receive channel active, the current consumption reduces to 2.3mA.

3.6.3 Transmitter

The peak global transmitter efficiency for the low frequency (LF) and high frequency (HF) channels was measured as 16.5% and 14.7% respectively, and occurs at an output power of 1.45mW. Higher efficiency is observed for the LF channel due to better alignment between the oscillator resonant tank and the power amplifier output filter. At an output power of 0dBm, the measured second- and third-harmonic distortion ratios (HD2 and HD3) for the LF channel are -38.4dB and -45.3dB below the carrier, respectively. For the HF channel, the measured HD2 and HD3 are -35.0dB and -45.7dB respectively.

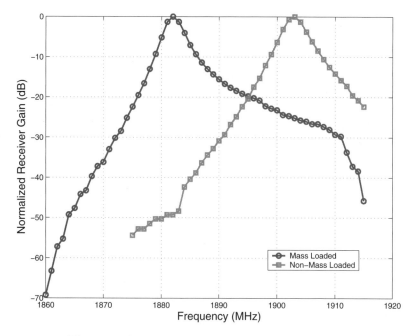

Fig. 3.15. Normalized receiver gain of both channels

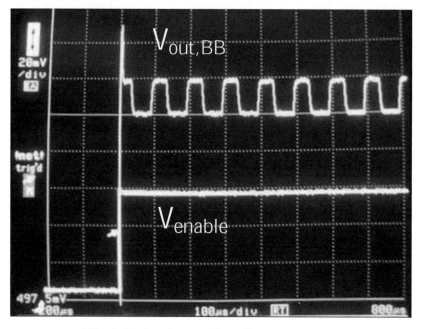

Fig. 3.16. Receiver enable-to-data start-up time

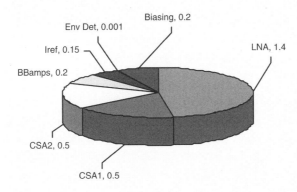

Fig. 3.17. Breakdown of the receiver current consumption (mA)

The start-up time of the transmitter determines the maximum achievable transmitter datarate. Figure 3.18 shows the measured start-up time of the transmitter to be approximately $1.5\mu s$. The transmitter achieved a maximum measured datarate of 160kbps.

3.7 Conclusions

The goal of this chapter was to explore the feasibility of placing the burden of filtering on a passive RF MEMS component instead of the traditional downconvert-and-filter method. This section described the design, implementation, and testing a low power transceiver using BAW resonators as the tuning elements. The benefits of this architecture are the avoidance of a crystal and PLL on the transmit side and the avoidance of any local oscillator on the receive side. The elimination of the quartz crystal allows the entire transceiver to be fabricated using thin-film techniques. The transceiver was fully functional and a robust 20m indoor wireless link was demonstrated. The architecture described in this chapter provides a modest sensitivity and is tolerant to frequency variations of the BAW resonators.

From Figure 3.17, we learned that a large majority of the power dissipation occurred in the amplification stages. Recall that a high RF gain was necessary to reduce the noise contribution of the self-mixing RF detectors. This is a problem common to all architectures that avoid traditional downconversion, and represents room for improvement in subsequent implementations. In the next chapter, this issue is addressed by exploring a super-regenerative receiver architecture.

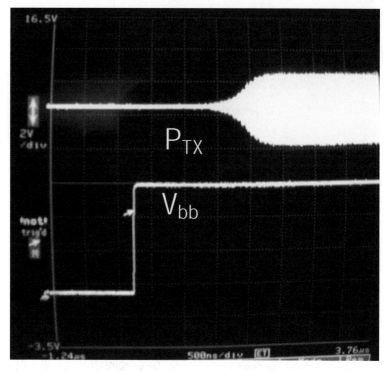

Fig. 3.18. Transmitter modulation start-up time

4

SUPER-REGENERATIVE RECEIVER DESIGN

So far, we have shown that RF MEMS technology can enable new circuit blocks and transceiver architectures. The next step is to push the boundaries and further decrease the size and power consumption of the transceiver. To continue this architecture exploration, a super-regenerative receiver architecture was evaluated. This chapter describes the design, implementation, and testing of a sub-mW BAW-based super-regenerative receiver. First, a brief history of this architecture is given.

4.1 History of the Super-Regenerative Receiver

The super-regenerative concept was first introduced by Armstrong in the 1920s [32]. Reasons for its success during the first half of the 20th century included a minimal number of required active devices (vacuum electron tubes), a high RF gain, and the ability to operate at high RF frequencies [1]. Typical RF circuit blocks use transistors biased at f_T values multiple times greater than the carrier frequency. The super-regenerative architecture is attractive because it allows receiver operation above the f_T of the RF devices. See Figure 4.1 for an example of a 1940s super-regenerative receiver [1]. This 500MHz receiver consists of a super-regenerative oscillator, an input coupling transformer, a tunable capacitor, and a detector tube. Manual adjustments are required to fine-tune the oscillator frequency and loop gain. Although the receiver is bulky and requires calibration, the few number of active devices and operation above device f_T is an impressive accomplishment. Antenna re-radiation, low spectral efficiency, distortion of analog modulation, and the advent of inexpensive transistors allowing super-heterodyne architectures rendered the super-regenerative receiver obsolete by the late 1950s. However, in this chapter it is shown that modern RF MEMS, CMOS technology, circuit design techniques, and digital communications solve many of the inherent problems with this architecture, providing very low power and high integration for sensor node applications.

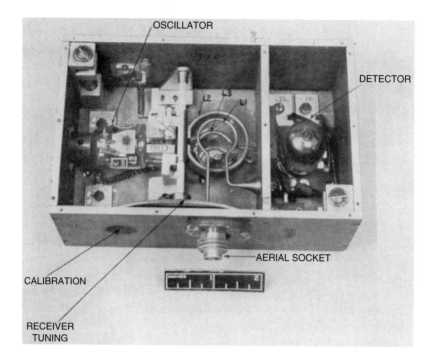

Fig. 4.1. Vintage 1940s two-tube super-regenerative detector [1]. Reprinted with permission of Cambridge Press

4.2 Motivation

As shown in Chapter 3, the tuned radio frequency architecture utilizes RF MEMS technologies to perform channel selection without the need for mixers or frequency synthesizers. However, a high RF gain is necessary due to the high noise injected by the self-mixing RF detection circuitry. A super-regenerative front-end provides extremely high RF amplification and narrowband filtering at low bias current levels. As shown in Figure 4.2, the heart of a super-regenerative detector is an RF oscillator with a time-varying loop gain. This block diagram consists of a passive matching network, an isolation amplifier, an amplifier with time-varying gain, and a bandpass positive feedback network forming an oscillator. The isolation amplifier between the antenna and the oscillator performs the following functions: it reduces RF leakage of the oscillation signal to the antenna, it provides an input match to the antenna via the passive matching network, and it injects the RF input signal current into the oscillator tank without adding significant loading to the oscillator. The time-varying nature of the loop gain is designed such that the oscillator transconductance periodically exceeds the critical g_m necessary to induce instability. Consequently, the oscillator periodically starts up and shuts off. The periodic shut-down of the oscillator is called "quenching".

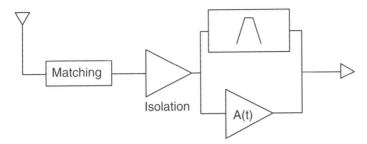

Fig. 4.2. Conceptual diagram of super-regenerative detection

The start-up time of an oscillator can be shown by Equation 4.1.

$$t_{rise} = \tau_{rise} \cdot ln\left[\frac{V_{osc}}{V_{initial}}\right] \tag{4.1}$$

where τ_{rise} is the time constant of the exponentially increasing oscillation envelope, V_{osc} is the zero-peak RF voltage of the saturated oscillator, and $V_{initial}$ is the zero-peak RF signal when the oscillator loop gain is unity (at the onset of oscillation). As this equation shows, the start-up time of the oscillator is exponentially dependent upon the initial voltage of the oscillator tank. This dependency provides the large gain attainable by the super-regenerative receiver.

There are two basic modes of operation: the logarithmic mode and the linear mode [1]. Waveforms (a), (b), and (c) in Figure 4.3 show the detector output in the linear mode, the output in the logarithmic mode, and the RF input signal, respectively. In the linear mode, the amplitude of oscillation is measured before the oscillator reaches saturation, providing a high signal-independent gain.

If the oscillator peak level is sampled before it reaches saturation, a signal-independent gain is realized. As shown in waveform (a), the sampled envelope is much larger in the presence of an RF input signal. Notice that, in this mode, the oscillator is disabled after the amplitude is measured. Thus, in the linear mode, the oscillator never reaches saturation. In the logarithmic mode, however, the oscillator is allowed to saturate during each cycle. Detection circuitry senses the area under the oscillation envelope, providing signal-dependent gain. Waveform (b) shows the increased area under the saturated oscillation envelope in the presence of an RF input, resulting from the decreased oscillator start-up time in this condition. Due to the severe fading anticipated in dense indoor sensor networks, a very wide dynamic range is required from the receiver. The logarithmic mode provides an inherent automatic gain control, making its use preferable for this application.

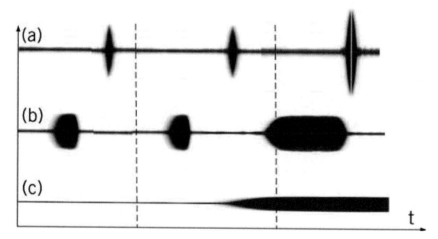

Fig. 4.3. Modes of super-regenerative operation. a) Linear, b) Logarithmic, c) RF Input

It should be noted that there are two very different methods to achieve a logarithmic-mode response. The first, as mentioned above, is the full on/off quenching of an oscillator by an external signal. This signal modulates, as a function of time, either the active device gain or the tank impedance. A second method of logarithmic mode excitation involves the use of a squegging oscillator. The term "squegging" refers to a parasitic mode that completely extinguishes the oscillation envelope before commencing oscillation again. Typically, this is caused by a DC bias point shift that occurs during the onset of oscillation, which gradually degrades the device transconductance and causes oscillation to cease periodically.

A super-regenerative oscillator may be operated in the squegging mode, which eliminates the need for an external quench signal. Detection is achieved by measuring the frequency of the squegging occurrences. In the presence of an RF signal, the oscillator starts up faster. Thus, the squegging frequency increases with the RF input level. It can be shown that detecting this frequency provides a logarithmic detection of the RF input, providing the same limitations and benefits as an externally quenched logarithmic mode detector. It should be noted that, in the squegging mode, no external quench signal is needed. The elimination of the quench signal is one main benefit of a squegging-mode receiver. However, for flexibility in the prototyping phase, an external quench signal is preferred for complete receiver characterization. Thus, for this implementation, an external quench logarithmic mode detector was designed. Ultimately, however, a squegging-mode receiver would be an efficient way to eliminate the need for an explicitly generated quench signal.

The potential of the super-regenerative receiver to generate large signal gain at very low bias currents makes it an attractive architecture for integrated ultra-low power wireless receivers.

4.3 Architecture

In contrast to traditional vacuum-tube implementations, the proposed architecture makes use of additional active devices to provide increased receiver performance. A block diagram for the proposed transceiver is shown in Figure 4.4.

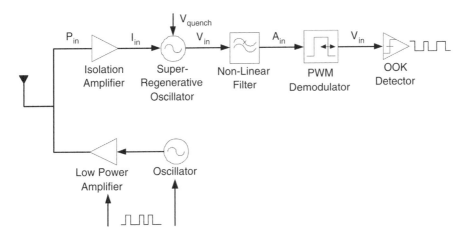

Fig. 4.4. Block diagram of proposed super-regenerative transceiver

The on-off keyed (OOK) transmitter consists of a BAW-referenced oscillator to provide a stable RF carrier and an antenna driver to provide efficient power gain. Digital bits are directly modulated onto the carrier by on/off cycling the transmitter. High transmitter efficiency is achieved through direct modulation and careful oscillator/driver co-design. At the receiving end, the RF input is matched to the 50Ω load presented by the isolation amplifier. The isolation amplifier converts the RF power to a current, injecting it into the detector oscillator while providing isolation between the oscillator and the antenna. The detector oscillator, whose time-varying tank impedance is cycled at 100kHz, samples the RF input signal as an initial condition for its growing exponential, modifying the start-up envelope. The receiver operates in the quenched logarithmic mode. Thus, the signal is sampled directly at RF by the detector oscillator, providing a large RF gain (>55dB) for low signal levels. A BAW resonator sets the free-running frequency of the detector oscillator. The envelope of this oscillation is detected by the non-linear filter, and its 100kHz sampling tone is removed by a pulse width demodulator, leaving a raw OOK analog signal. The baseband signal can be readily detected with an analog matched filter, an A/D with digital baseband, or a one bit slicer with an appropriately placed threshold.

4.4 Analysis

This section will analytically describe the operation of super-regeneration in the logarithmic mode. We will begin with a qualitative explanation of the receiver operation. An analysis of the super-regenerative gain and bandwidth follows. Finally, limitations of the quench frequency are derived.

4.4.1 Operation

The RF signal is coupled into the oscillator tank from the antenna. It is then discretely sampled by the time-varying detector oscillator. Since this is a discrete-time sampled process, the quench rate must be at least twice the highest frequency component of the baseband signal. This quench tone must be filtered from the data signal to avoid corrupting the baseband signal. Thus, the oversampling ratio is determined by the complexity of this filter. The datarate is ultimately limited by the bandwidth of this filter. However, the quench frequency cannot be arbitrarily increased to achieve a higher datarate. The frequency of the quench signal is limited by the time varying growing exponential time constant of the oscillator, τ_{det}. For a high Q resonant structure, such as a BAW resonator, τ_{det} is relatively long (greater than 100ns). To allow a higher quench frequency, a short duty-cycled quench signal (e.g. 10%) can be utilized. In this implementation such a short duty cycle is possible due to the very rapid quenching of the oscillator tank by the shunting transistor. A majority of the quench cycle can be dedicated to oscillator start-up and detection instead of quenching.

4.4.2 Super-Regenerative Gain

The following is a derivation of the gain of a super-regenerative receiver in the logarithmic mode. First, let us consider the startup time of an oscillator (the time from enabling the oscillator until it reaches its saturation voltage V_{osc}) as shown in Equation 4.2.

$$t_{rise,noise} = \tau_{rise} \cdot ln\left[\frac{V_{osc}}{\sqrt{v_n^2}}\right] \tag{4.2}$$

Notice that this is the same as Equation 4.1 except that the initial oscillator signal is the thermal noise of the oscillator tank and active devices. Thus, we predict that the oscillator will actually reach saturation in a different amount of time each time it starts up. This is indeed the case, as shown by the measured histogram of oscillator startup transients in Figure 4.5.

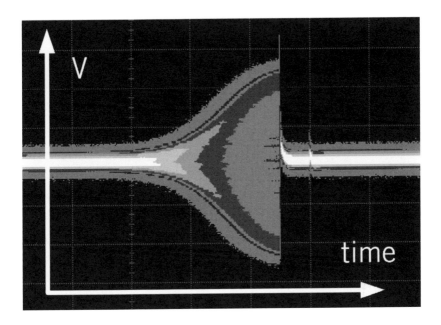

Fig. 4.5. Startup of RF oscillator in the presence of noise

This histogram was created by superimposing many startup traces, allowing a visualization of the uniformity of the startup time. The histogram shows that the startup time of the oscillator is poorly defined (fuzzy light colored leading edge) since it is directly related to the instantaneous noise in the tank at as the oscillator loop gain exceeds unity. Observing that the startup time is very sensitive to low-level signals in the oscillator tank, it becomes clear that this mechanism can be used to amplify desired signals as well. As the input signal current is injected into the oscillator tank, it is converted to a voltage V_{rf}. For large RF signals, the startup time of the oscillator is given by Equation 4.3.

$$t_{rise,signal} = \tau_{rise} \cdot ln\left[\frac{V_{osc}}{V_{rf}}\right] \qquad (4.3)$$

As shown in [1], one can quantify the signal to noise ratio as the area between the two exponential Equations 4.2 and 4.3. This is given by Equation 4.4.

$$\Delta_{area} = V_{rf} \int_0^{t_{rise,sig}} e^{\frac{t}{\tau}} dt + V_{osc}(t_{rise,noise} - t_{rise,signal}) - \sqrt{\overline{v_n^2}} \int_0^{t_{rise,noise}} e^{\frac{t}{\tau}} dt$$

(4.4)

To extract the modulated data from this signal, a running average of the oscillator envelope is taken, yielding V_{bb}. V_{bb} represents the demodulated output voltage in response to an RF input. In this analysis, DC is the quench waveform duty cycle and f_q is the quench frequency. Equation 4.5 gives the demodulated output assuming a high receiver SNR.

$$V_{bb} = V_{osc}\left[DC - \frac{\tau}{T_q}\ln\left(\frac{V_{osc}}{2V_{rf}}\right)\right]$$

(4.5)

V_{osc} is the RF oscillator zero-peak swing, $T_q = \frac{1}{f_q}$, V_{rf} is the RF resulting from the input RF power, and τ is the oscillator start-up time constant. The receiver gain, therefore, is given by Equation 4.6.

$$Gain = \frac{V_{bb}}{V_{rf}} = \frac{V_{osc}}{V_{rf}}\left[DC - \frac{\tau}{T_q}\ln\left(\frac{V_{osc}}{2V_{rf}}\right)\right]$$

(4.6)

Notice that the receiver gain is approximately a linear function of the input signal amplitude. Thus, from a system perspective, the receiver can be modeled quite accurately as a linear amplifier followed by a logarithmic detector. This property will be revisited in the measurements section of this Chapter.

4.4.3 Super-Regenerative Bandwidth

As shown in the block diagram in Figure 4.2, the receiver is tuned by a bandpass filter with bandwidth f_{FB}, which sets the free-running oscillator frequency. For this analysis, we will assume that this bandpass filter block has a second-order, high Q resonance.

By applying positive feedback to the system with a gain $A(t)$, the poles move closer to the $j\omega$ axis, increasing the Q of the system. This mode is called the *regenerative mode*. In this mode, a gain enhancement is achieved since the tank impedance increases. In addition, the receiver bandwidth decreases, resulting from the Q enhancement of the positive feedback. As one may expect, such a system is prone to oscillation and requires frequent calibration. However, as observed by Armstrong, periodically allowing oscillation can provide the same benefits as the regenerative mode with increased robustness. This section derives the frequency response of this mode, called the

$super - regenerative$ mode. For the bandwidth analysis, we will limit our discussion to the logarithmic mode of super-regeneration.

Let us define g_{tank} as the effective conductance of the tank. The bandwidth of the tank is then:

$$BW_{-3dB} = \frac{f_o}{Q_{tank}} = \frac{g_{tank}}{k} \qquad (4.7)$$

Where k is a constant that will be subsequently canceled. Let $g_{active}(t)$ represent the time-varying negative conductance presented to the tank by the active devices. The total conductance is given by $g_t(t) = g_{tank} - g_{active}(t)$. When $g_t(t)$ becomes negative, the exponential oscillator start-up ensues. The RF input signal plus noise is sampled as an initial condition for this growing exponential. The magnitude of g_t thus effects the bandwidth of the receiver, as will now be shown.

During super-regeneration, the bandwidth is modified by the new pole locations to yield Equation 4.8.

$$BW_{-3dB} = \frac{f_o}{Q_{tank}} \frac{(g_{tank} - g_{active})}{g_{tank}} = \frac{f_o}{Q_{tank}} \frac{(g_t(t))}{g_{tank}} \qquad (4.8)$$

During operation, this appears as two coupled tanks, giving the following -6dB receiver bandwidth [1].

$$BW = \frac{g_{tank}g_t}{(g_{tank} + g_t)k} \qquad (4.9)$$

Where k is the same constant given in Equation 4.7. Thus, we can normalize the super-regenerative bandwidth to that of the tank resonator only. This would provide the amount of bandwidth narrowing that occurs during the super-regeneration process.

$$\frac{BW_{SR}}{BW_{tank}} = \frac{g_t}{g_{tank} + g_t} = R \qquad (4.10)$$

Where we will define R as our bandwidth narrowing factor. Equation 4.10 indicates that, if a great deal of bandwidth narrowing is needed, the transconductance g_t must be very tightly controlled. For a high bandwidth narrowing factor R, the tank conductance must be greatly reduced in the super-regenerative mode, causing the high precision requirements on the transconductance. To quantify the sensitivity, we must relate the bandwidth to the oscillator transconductance. Here we will consider the tank conductance only during super-regeneration, removing the time-varying nature of the tank and using the magnitude of the oscillator transconductance g_m for simplicity.

$$g_t = |g_m| - g_{tank} \qquad (4.11)$$

The initial loop gain of the oscillator is given by Equation 4.12.

$$A_{initial} = \frac{g_m}{g_{tank}} \tag{4.12}$$

We can now write the initial loop gain of the super-regenerative oscillator in terms of the desired bandwidth narrowing.

$$A_{initial} = 1 + \frac{R}{1-R} \tag{4.13}$$

For example, assuming an on-chip LC tank at 2GHz with $L = 10nH$, $C = 633.2fF$, and $Q = 10$ would yield a bandwidth of 200MHz and $g_{tank} = 795.8\mu S$. A reasonable desired receiver bandwidth is 500kHz. To provide a 500kHz super-regenerative bandwidth, the tank Q must be narrowed by almost three orders of magnitude ($R = 400$). To achieve this, Equation 4.10 dictates that the total transconductance during super-regeneration must be $g_t = (-2.5x10^{-3})g_{tank}$. Conceptually, to create a high Q system out of a low Q tank, the poles must be placed very near the $j\omega$ axis, necessitating very tight control on the oscillator transconductance. Equation 4.13 indicates that the initial loop gain of the oscillator must be controlled to $A_{initial} = 1.0025$. In a super-regenerative implementation, this requirement manifests itself as the need to have a precisely set quench amplitude. This is traditionally accomplished through hand-tuning or complex amplitude control loops [33].

In contrast, if one starts with a high Q tank, the transconductance precision requirements are greatly reduced. Now assume a tank consisting of a 2GHz BAW resonator with $Q = 1000$. This yields a tank bandwidth of 2MHz. To achieve a 500kHz receiver bandwidth, the tank must be narrowed by only 4x ($R = 4$). Equation 4.10 now indicates that the total transconductance during super-regeneration must be $g_t = (-0.333)g_{tank}$. This corresponds to an initial loop gain of approximately $A_{initial} = 1.3$, which is much easier to achieve in practice.

The sensitivity of the receiver bandwidth to transconductance variations is also extremely important. For a tank with $Q = 10$, a 5% increase in oscillator transconductance causes a 1895.2% increase in the receiver bandwidth. A 5% decrease in transconductance would eliminate oscillation altogether, disabling the receiver operation. For a high tank ($Q = 1000$), a 5% transconductance variation causes a 14% receiver bandwidth variation, allowing drastically improved robustness in practice. In addition, incremental reductions in transconductance would not eliminate receiver operation. As another specific example, it is interesting to calculate the necessary transconductance accuracy for a 20% receiver bandwidth accuracy. For a $Q = 1000$ tank, the transconductance accuracy must be greater than 7%. For a $Q = 10$ tank, the transconductance accuracy must be better than $5x10^{-5}\%$. This highlights the very high sensitivity to bias current and transconductance variations that occur when trying to narrow the bandwidth of a low Q tank through super-regeneration.

In this work, a high Q tank is used with a low R factor. Thus, a digital square wave quench signal is used and modest oscillator bias current variations have minimal impact on the receiver bandwidth. This will be shown experimentally in Section 4.7.

4.4.4 Quench Frequency Limitations

The receiver quench frequency is important because it determines the maximum receiver datarate. Because the receiver performs a discrete sampling process, $f_{bb} < \frac{f_q}{2}$. In addition to the sampling rate, the oscillator must decay to below the receiver noise level before the next quench cycle starts. If this condition is not met, the initial condition of the oscillator start-up will be the previous cycle's oscillation and not the desired input RF signal. Equation 4.14 shows the oscillation voltage as a function of time during the decay phase.

$$V_{decay}(t) = V_{osc} \cdot e^{\frac{-t}{\tau_{decay}}} \qquad (4.14)$$

To reduce the oscillation to below the oscillator noise floor, the following condition must be met:

$$V_{osc} \cdot e^{\frac{-t}{\tau_{decay}}} < \sqrt{v_n^2} \qquad (4.15)$$

were $\sqrt{v_n^2}$ is the oscillator noise voltage. Thus, the damping time of the quench phase is given by Equation 4.16.

$$t_{damp} > \tau_{decay} \cdot ln\frac{V_{osc}}{\sqrt{v_n^2}} \qquad (4.16)$$

Additionally, because the receiver input signal is discretely sampled by the RF oscillator, noise aliasing occurs. The excess noise factor due to noise folding is given by Equation 4.17.

$$N_{excess} = 10 \cdot log\left[\frac{BW_{noise}}{f_{quench}}\right] \qquad (4.17)$$

Where BW_{noise} is the effective brick-wall RF input bandwidth. To reduce noise folding, f_{quench} should be maximized. By substituting for the maximum allowable quench frequency as defined above, and the noise bandwidth, Equation 4.18 is obtained.

$$N_{excess} = 10 \cdot log\left[\left(\frac{f_0}{Q}\right)\frac{2Qln(\frac{V_{osc}}{\sqrt{v_n^2}})}{f_0\pi}\right] \qquad (4.18)$$

This relationship can be simplified to Equation 4.19.

$$N_{excess} = 10 \cdot log\left[\left(\frac{2}{\pi}\right) \cdot ln\left[\frac{V_{osc}}{\sqrt{\overline{v_n^2}}}\right]\right] \quad\quad (4.19)$$

In summary, the discrete-time nature of the RF sampling process causes aliasing to take place, which increases the noise figure of the receiver. For typical values, the excess noise may exceed 5dB. Notice that this noise factor is independent of the resonator Q. The folding factor may be reduced by modifying the quench waveform to provide higher resonator damping factors, allowing a higher quench frequency and less noise folding. Modern circuit technologies allow the use of an additional active device that acts as a switch to dissipate the RF energy quickly after each quench period. This allows a near-instantaneous oscillator quench, reducing the effects of aliased noise on the receiver noise figure.

4.5 LNA/Oscillator Design

The main considerations in the isolation amplifier design were the following: fully-integrated input matching network achieving an S_{11} of $<$10dB at 1.9GHz, very low current operation, and sufficient reverse isolation of the detector oscillator signal. Because the current consumption of this stage is set by the detector oscillator, a PMOS amplifier topology was chosen for the isolation amplifier due to the higher transconductances of the NMOS devices. For an inductor degenerated amplifier, the input match occurs at the resonance of the gate and source inductors and the gate-to-source device capacitance (C_{gs}) of the input transistor. This input device was sized for weak inversion operation. While this reduces the f_T of the device, it allows a reasonable gate inductor size (10nH), facilitating full integration of the input matching network. The cascode device sizing is a tradeoff between reverse isolation, headroom, and total amplifier transconductance.

The 1V, 400μA receiver front-end schematic is shown in Figure 4.6.

The isolation amplifier comprises an inductively degenerated PMOS LNA with two on-chip inductors, yielding a fully integrated matching network. The most power-hungry components in the receiver - the isolation amplifier and detector oscillator - share their bias current, thereby effectively halving the current consumption of the receiver. The detector oscillator is cycled by the quench signal (V_{quench}), which creates a time varying tank impedance, periodically dissipating the RF energy stored in the BAW resonator through a shunting transistor. The shunting transistor must be sized wide enough to quickly quench oscillation, but small enough to avoid loading the tank with unnecessary capacitance. A 50Ω on-resistance is sufficient, yielding the addition of very low parasitic capacitance. The power consumed in driving the quench transistor must be absorbed into the receiver power budget. For a 300fF capacitance (including pad and wiring capacitance), this power dissipation is

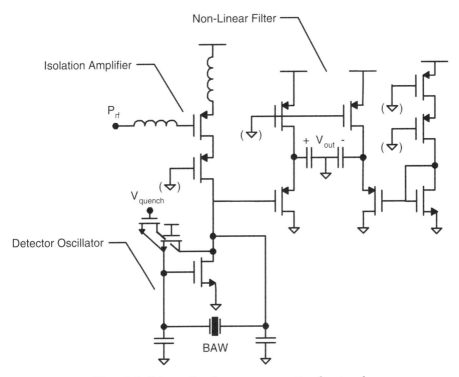

Fig. 4.6. Schematic of super-regenerative front-end

approximately 30nW. The shape of this impedance waveform allows for the tuning of the receiver gain and bandwidth properties [34]. A square wave (10% duty cycle) impedance waveform allows simple and adaptable quench generation. Because the data symbols are oversampled, the exact quench frequency and phase is not crucial, and may be readily supplied from a digital control block. The high Q nature of the resonant BAW structure, providing a relatively long oscillator time constant and narrow intrinsic bandwidth, relaxes the need for precise control over the oscillator transconductance. Fine Tx/Rx frequency alignment can be achieved with relatively large, binary-weighted capacitor arrays due to the high Q resonator.

A weak-inversion PMOS non-linear filtering stage is DC-coupled to the oscillator. This stage consists of a PMOS source follower with a relatively low pole frequency (1MHz). The 2GHz carrier is attenuated, but the non-linearity of the PMOS transistor creates a DC component proportional to the envelope of oscillation. By using a DC coupled topology, the non-linear bias point shift of the oscillator NMOS transistor adds to the PMOS non-linear filter, thereby increasing the signal level to the pulse-width demodulator.

A replica stage (with 20x current division relative to the super-regenerative core) was used to provide a pseudo-differential output of the non-linear filter.

4.6 Additional Circuitry

In addition to the RF front-end, the prototype integrated a two-stage RF buffer, a non-linear filter, and a baseband buffer. The RF buffer allows access to the core of the super-regenerative oscillator. Although not needed for normal receiver operation, access to this node is useful for diagnostic purposes. The buffer is self-biased, two-stage, DC coupled, and common-mode insensitive. It presents a 50fF capacitive load to the super-regenerative oscillator and drives a 50Ω, 1pF test equipment load. The RF buffer was included for diagnostic purposes, and allows full characterization of the super-regenerative oscillator operating points.

4.7 Experimental Results

This section documents the experimental verification process for the super-regenerative prototype.

4.7.1 Board Design

A custom printed circuit board was designed, fabricated, and populated. The chip and accompanying resonators were assembled with chip-on-board (COB) technology; no special assembly techniques were used for the resonator mounting. The prototype receiver implementation is shown in Figure 4.7. Board trace lengths were minimized in order to reduce interconnect parasitics. Finally, by-pass capacitors were used on supply lines to reduce noise.

4.7.2 Testing Methodology

For receiver testing, a modulated RF signal generator was used to provide a test signal. An *HP3764A* Bit Error Rate Tester (BERT) was used to provide real-time BER measurements. A pseudorandom sequence of length 2^{23}-1 (8,388,607) was used as the data pattern, providing a non-repeating pattern for 27.9 minutes at 5kbps. Assuming 10 error hits are needed for an accurate BER measurement, 20 seconds of measuring time is needed for a 10^{-4} BER at 5kbps. For a BER measurement of 10^{-5} and 10^{-6}, 200 and 2000 seconds (33.3 minutes) are needed, respectively. Thus, measurement resolution is limited to a BER of approximately 10e-6.

Fig. 4.7. Die photograph of the super-regenerative receiver

4.7.3 Results

The combined active circuit and inductor area is less than 1mm^2. The accompanying 1.9GHz BAW resonator is wirebonded directly to the CMOS die to eliminate board parasitics. The measured S_{11} of the fully-integrated matching network is shown in Figure 4.8. The plot shows two distinct input matching curves. Direct chip-level probing of the circuit yields an input match of -27.5dB at 1.894GHz. This is an extremely good match, considering this was a first revision circuit using a fully-integrated matching network. Placement of the chip on a COB test board provides a -25dB input match at 1.7GHz. The reduction in input match frequency is due to the bondwire inductance and the parasitic board capacitance. Because COB board placement is a likely permanent solution, this mismatch in input frequency was corrected in a revised design. The measured eye diagram of the detector oscillator RF signal in the presence of a -80dBm OOK input is shown in Figure 4.9, illustrating the variation in oscillator startup time for "1" and "0" symbols. The measured sensitivity for a bit error rate (BER) of 10^{-3} is -100.5dBm at 5kbps,

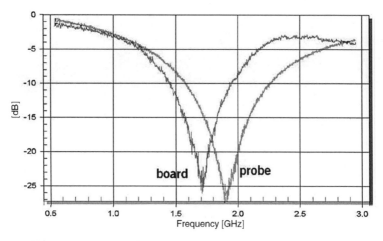

Fig. 4.8. Measured S_{11} magnitude of super-regenerative receiver

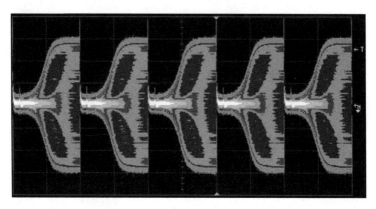

Fig. 4.9. Super-regenerative eye diagram in the presence of a -80dBm signal

displaying negligible degradation over a supply voltage range from 0.9-1.3V. For a quench frequency of 100kHz and a 10% quench duty cycle, the RF small signal bandwidth of the receiver is 0.5MHz. The measured gain response of the receiver, displayed in Figure 4.10, shows the signal-dependent gain characteristic of super-regenerative detectors operated in the saturated oscillator mode.

In a vintage super-regenerative transceiver, the logarithmic mode was avoided since unacceptable distortion to analog modulation was introduced by this transfer function. However, for digital OOK communication, the logarithmic gain response provides the receiver with an inherent automatic gain control (AGC), easing the dynamic range requirements of the baseband circuits. AGC is a desirable trait in transceivers for indoor wireless sensor networks

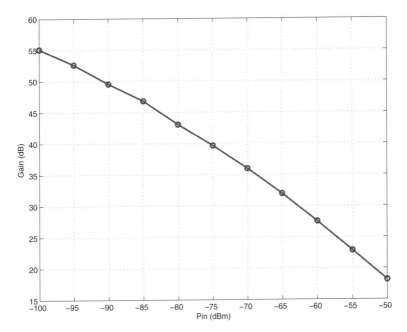

Fig. 4.10. Measured signal-dependent gain response of super-regenerative receiver

which must accommodate large input power variations due to deep fading and interferers.

To ascertain the performance of the receiver over a range of supply voltages, V_{dd} was swept while monitoring the BER at the output of the demodulator. The experimental data is shown in Figure 4.11. Negligible degradation in BER occurs from 0.9V-1.3V. For a V_{dd} of 0.8V, there is a reduction in sensitivity of approximately 0.5dB, a result of the input PMOS in the isolation amplifier being forced out of saturation by the self-biased cascode device. This sensitivity data was taken with a 20kHz demodulator bandwidth, reducing the sensitivity slightly. The receiver is thus expected to operate over the entire battery life of a typical cell.

The radiation leakage of oscillator power back to the antenna (receiver re-radiation) is an important metric in super-regenerative receivers. Historically, this phenomenon has been a problematic cause of interference when the input signal is transformer-coupled into the oscillator [1]. The use of a cascoded isolation amplifier limits the measured re-radiation at the antenna port to -69dBm[1].

---

[1] Modulated with a 100kHz quench tone

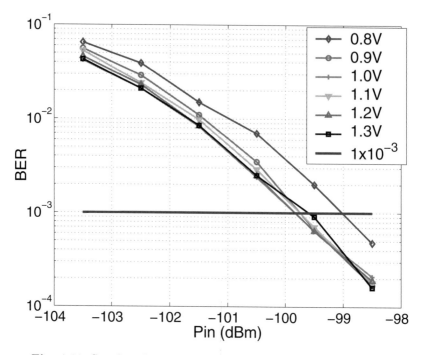

Fig. 4.11. Supply voltage dependence of super-regenerative receiver

4.8 Discussion

This chapter described the design and implementation of a super-regenerative receiver using BAW resonators. The main benefits of a super-regenerative architecture are:

- Operation at carrier frequencies above the f_T of the active devices, allowing subthreshold device operation and/or high carrier frequencies.
- Fabrication using entirely thin-film processes. This is made possible by using thin-film RF MEMS resonant structures and integrated passive elements.
- Very high receiver sensitivity due to the high RF gain generated in the super-regeneration process and the relatively small signal bandwidth, allowing reduced transmitted power.

There are also potential liabilities of using a super-regenerative architecture. First, traditional super-regenerative receivers require a hand-tuned RF frequency reference. Since the local oscillator is periodically quenched, a typical PLL referenced to a quartz crystal oscillator cannot be used. One solution is to periodically lock the super-regenerative oscillator to a PLL [35]. In this work, we eliminate the need for a PLL by utilizing a high frequency MEMS reference.

Another limitation is the selectivity of the receiver. Although it has been shown that the bandwidth of the receiver is narrowed significantly beyond what is achievable with the resonant structure itself, super-regenerative receivers historically have suffered from poor interferer tolerance. This property will be explored in more detail in the next chapter.

Table 4.1 shows the measured performance characteristics of the prototype super-regenerative receiver.

Table 4.1. Super-Regenerative Receiver Performance Summary

Parameter	Performance
Power Consumption	$400\mu W$
V_{dd}	1V
Datarate	5-10kbps
Sensitivity (5kbps)	-100.5dBm
Quench Frequency	100kHz
S_{11} (probe)	-27.5dB
S_{11} (COB)	-11dB

A subsequent version was designed that integrated the pulse width demodulator, the transmitter, and all biasing circuitry. This effort is described in detail in the next chapter.

5

FULLY INTEGRATED
SUPER-REGENERATIVE TRANSCEIVER

Chapter 4 described the analysis, design, and testing of a super-regenerative prototype receiver. That work was the basis for the fully integrated super-regenerative transceiver that will be described in this chapter, combining the design strategies developed during the first few chapters. The goals of this design were as follows:

- Minimize the die area of a fully integrated, sub-mW transceiver;
- Achieve full integration: no crystals, inductors, or biasing resistors;
- Integrate programmable transmit and receive RF frequencies;
- Integrate programmable baseband gain and frequency; and
- Utilize a serial interface to minimize digital control pads.

This chapter will describe the design and implementation of a 1.9mm^2, $400\mu\text{W}$ super-regenerative receiver and integrated transmitter [36].

5.1 Architecture

Referring back to Figure 4.2 shows the basic architecture used here. The pulse width demodulator and the transmitter are now integrated on the chip. In addition, a digital serial interface was added to allow control over all receiver operation. The front end bias current, front end RF frequency, and baseband filter cutoff frequency and gain are all programmable through this interface.

5.2 RF Front-End Circuit Design

Recall Figure 4.6, showing the front-end circuit schematic. In the fully integrated chip, the layout of the isolation amplifier was modified to reduce the substrate coupling from the super-regenerative oscillator back to the antenna. A guard ring separates the oscillator from the isolation amplifier in order to suppress leakage. A tradeoff exists between the substrate coupling suppression between the oscillator and amplifier and the parasitics added by the critical coupling node between the two blocks. Additionally, the on-chip input matching network was modified to accommodate the anticipated COB assembly parasitic impedances (refer to Figure 4.8).

To eliminate all off-chip biasing circuitry, a digitally programmable on-chip bias circuit was implemented. Current source digital-to-analog converters (DACs), controlled by the SPI interface, were used to provide a variable bias current to the RF front-end. See Figure 5.1 for the DAC schematic.

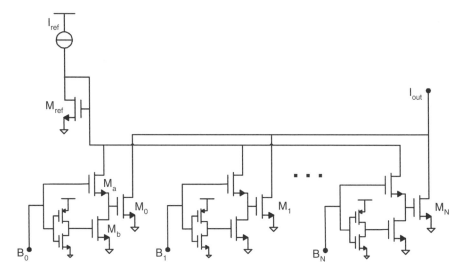

Fig. 5.1. Simplified schematic of the current source DAC

The reference current I_{ref} is generated on-chip. Transistors M_0 through M_N are matched binary weighted aspect ratio devices. The gate voltage of transistor M_{ref} is switched to the array via transistor M_a. M_b is driven with an inverted binary code to ensure positive pull-down of all current source gates. For the super-regenerative oscillator biasing, a 6-bit DAC was designed, providing bias currents from 0 to $630\mu A$ with a resolution of $10\mu A$.

The DAC layout consumed $(80 \times 40)\mu m^2$, including 4pF of decoupling capacitance and a biasing resistor. Dummy cells and randomized unit cell placement was utilized in the layout of the binary weighted array to ensure good DAC linearity. The layout is shown in Figure 5.2.

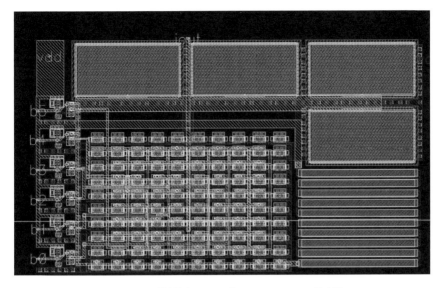

Fig. 5.2. CAD layout of current source DAC

5.3 Pulse Width Demodulator

One crucial block of the super-regenerative receiver is the pulse width demodulator. This circuit is responsible for separating the quench tone from the desired signal. To accommodate variable quench frequencies and datarates, a programmable filter is necessary. In addition, in order to accommodate various baseband detection methods, a programmable gain is desirable. The following section describes the filter specification and realization.

5.3.1 Filter Specifications

The signal entering the filter is an envelope of the super-regenerative oscillator output. It consists of a pulse train modulated at the quench frequency (approximately 100kHz). The duty cycle of this pulse train is dependent upon the RF signal strength at the antenna. Mathematically, the demodulated signal is extracted by taking the average of this pulse train. In practice, this can be achieved by a multi-order low pass filter. This filter must separate the baseband signal from the strong quench frequency tone. The quench tone appears as ripple at the filter output. Figure 5.3 shows the simulated input and output waveforms of the pulse width demodulator. A 100mV pulse train at 100kHz simulates the output pulse train of the detector oscillator. The two overlaid curves are the ideal 3rd order butterworth filtering and the simulated output of the full active filter implementation, which closely match. The non-ideal output waveform is the result of using non-linearized transconductor

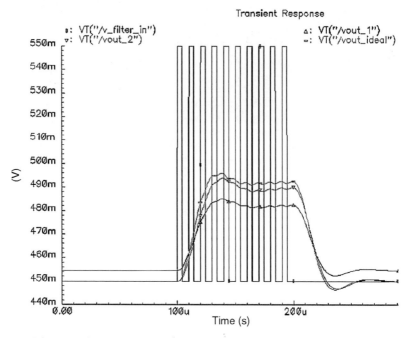

Fig. 5.3. Simulated input/output waveforms of the demodulator

elements. The filter circuit design will be described in detail in the next section. To ensure that the receiver sensitivity is not compromised by the filter, this ripple must be reduced below the noise floor of the receiver, corresponding to an attenuation of approximately 40dB at 100kHz. It should be noted that the filter also performs anti-aliasing of a subsequent A/D converter.

The design of the filter must trade off filter complexity and datarate. As the filter order increases, the filter cutoff frequency can be placed closer to the quench frequency, allowing higher datarates. System-level Matlab simulations were used to identify the optimal filter order. Figure 5.4 shows the simulation results for a 100kHz quench frequency at a 20kHz filter cutoff frequency. The simulation shows demodulation of 10kbps OOK data and the resulting quench ripple. The duty cycle of the pulse train is varied from 15% to 30%, simulating the receiver operation under low SNR conditions.

A 3rd order Butterworth filter allows sufficient quench signal rejection for this cutoff frequency to allow reception at the receiver RF minimum detectable signal level. For higher signal levels, additional ripple can be tolerated. Thus, the filter bandwidth should be dynamically adjustable to accommodate variable datarates over varying levels of SNR. Additionally, to ease the implementation requirements on the baseband detector, digitally programmable gain

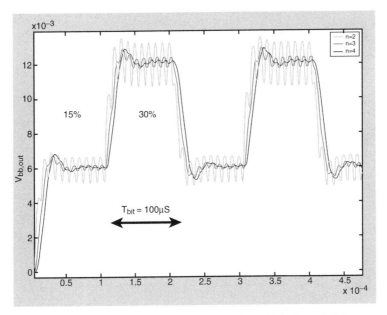

Fig. 5.4. System-level simulation of pulse-width demodulator

should be included in the filter path. These requirements will now be used to synthesize and design the baseband pulse-width demodulator filter.

5.3.2 Filter Synthesis and Design

The 3rd order Butterworth filter was synthesized for a nominal cutoff frequency of 20kHz. A terminated, lumped C-L-C filter ladder was chosen, yielding $C_1 = 795.7$pF, $L = 0.159$H, and $C_2 = 795.7$pF [37]. Conservative noise scaling was used to render the filter noise contribution negligible in relation to the quench ripple and receiver noise. Scaling yielded $C_1 = 39.78$pF, $L = 3.18$H, and $C_2 = 39.78$pF. The large-value inductor was synthesized with a 3-OTA, one capacitor gyrator. The result of the synthesis is the filter ladder shown in Figure 5.5.

A folded-cascode structure was chosen for good low voltage performance. The input common mode range and output voltage range are large, allowing the potential for 1V operation. The filter transconductor schematic is shown in Figure 5.6.

Although modern processes necessitate low supply voltages which reduces the circuit headroom, multiple threshold voltages are typically provided. These transistors can be utilized to facilitate low voltage analog design without sacrificing performance. A combination of high V_t and low V_t transistors was used to achieve optimal V_{ds} values in the output cascode stacks in order to maximize the output swing. The input transconductors are biased in

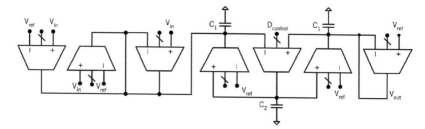

Fig. 5.5. Pulse width demodulator filter ladder

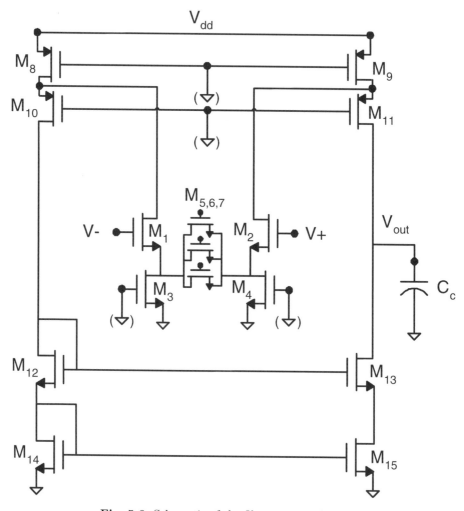

Fig. 5.6. Schematic of the filter transconductor

deep subthreshold at 800nA with a $\frac{40\mu m}{0.5\mu m}$ aspect ratio (IC=0.01). Linearization is accomplished through a degenerated split-source input differential pair. The sources of M_1 and M_2 are coupled with the trioded transistors $M_{5,6,7}$. Frequency tuning is achieved by digital control of the gates of these transistors, for which three bits of control are used. The same OTA is used as a termination resistor, yielding digitally adjustable gain control. Each OTA consumes $3.2\mu W$ at 1V. The OTA achieves an open-loop DC gain of 60dB over a 250-700mV output swing. The minimum phase margin at the maximum transconductance setting is 65°. To reduce the OTA offset voltage, a common centroid structure was used on the input pair. See Figure 5.7 for the OTA layout. Furthermore, large area transistors (greater than $100\mu m^2$) were

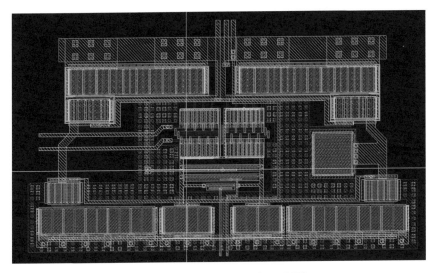

Fig. 5.7. Layout of the filter OTA

used for $M_{8,9,14,15}$ to reduce the flicker noise contribution. The filter utilizes a total of 10 OTAs, yielding a total power consumption of $32\mu W$. The filter bandwidth is digitally programmable from 9kHz to 30kHz. Programmable gain from from 6dB to 22dB is achieved. The acceptable filter common mode input range is 300mV to 600mV, and can tolerate a +/-50mV input offset without saturating. The layout of the filter is crucial to obtain good matching between OTAs and filter capacitors. The filter layout is shown in Figure 5.8.

The ten OTAs, filter capacitors, and a bias generator are clearly shown in the plot. A compact layout is essential, because the area of the filter occupies a sizeable percentage of the total receiver die area. The completed filter consumes $(350 \times 450)\mu m^2$.

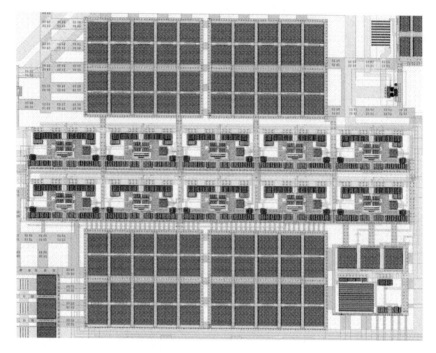

Fig. 5.8. Layout of the pulse width demodulator

5.4 Capacitance Tuning

The frequency of the detector oscillator (and, thus, the absolute receiver frequency) is programmable by a binary weighted capacitor array. Due to the high Q resonant structure used, fine frequency steps may be achieved using relatively large unit capacitors. This was shown in Equations 2.2 and 2.3. The bandwidth of the receiver is 500kHz, setting the limit for the frequency tuning resolution. A 3-bit capacitive array was used to provide 500kHz tuning steps over a total frequency range of approximately 3.5MHz, allowing a Tx/Rx alignment that is insensitive to small changes in capacitance. Equation 2.2 provides the frequency sensitivity to capacitive variation. The theoretical frequency shift of the BAW parallel resonant frequency to changes in capacitance is given by Equation 5.1.

$$\frac{\delta f_p}{\delta C_T} \simeq f_{series} \frac{-C_x}{2C_T^2} \tag{5.1}$$

For 500kHz steps, a 60fF least-significant bit (LSB) capacitance is needed. The switch transistor design entails an optimization of the tradeoff between series resistance and un-switched capacitance, both of which reduce the oscillator performance. The result of the optimization was a $\frac{20\mu m}{0.13\mu m}$ NMOS switch aspect ratio and a 80fF unit cell LSB capacitance was used, providing a switched capacitance of 60fF and a series switch resistance of approximately 25Ω at a 1V switch voltage, yielding a switched capacitor quality factor of approximately 50.

Two matched capacitor arrays were used, one on each side of the BAW resonator. Figure 5.9 shows the implementation of the capacitor array.

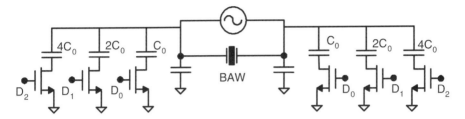

Fig. 5.9. Schematic of binary weighted switched capacitor tuning array

It should be noted that, due to Miller multiplication, a factor of two reduction in the necessary capacitance could be achieved by using one array shunting the BAW resonator. However, it is advantageous to use two non-Miller multiplied arrays because the capacitance on either side of the resonator forms the feedback network. To optimize loop stability, any additional capacitance should be added to these feedback capacitors, not the BAW shunting capacitance. Capacitive tuning allows compensation for resonator fabrication uncertainty, a means of temperature compensation, and the opportunity for both frequency hopping and interferer avoidance.

5.5 Chip Implementation

The goal of the chip floorplanning was to reduce the die area, ensure RF signal integrity, and integrate as much functionality as possible on-chip. To reduce the die size, a three pin serially programmable interface (SPI) was implemented, allowing control over the transceiver control bits. All biasing was generated internally and controlled digitally, eliminating external references and trimming. The RF receive path consumes an area of 0.5mm^2, and was isolated from the transmitter and SPI interface to minimize coupling. See Figure 5.10 for a photograph of the assembled chip.

The left side of the chip (RF front-end) is the "quiet" side, and the right hand side (transmitter) is the "noisy" side. As the signal passes from left-to-right, from the super-regenerative oscillator to the non-linear filter and through the pulse-width demodulator, the signal frequency decreases, making the signal less susceptible to coupling. An output buffer located in close

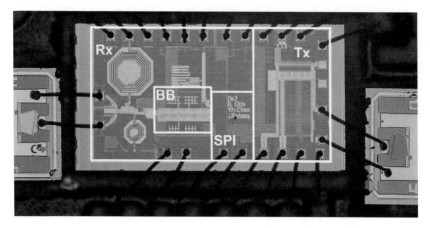

Fig. 5.10. Micrograph of the assembled integrated transceiver

proximity to the demodulator drives the signal to the output pads. The buffer provides drive capability for a 2kΩ, 20pF off-chip load.

The transmit and receive resonators are clearly visible on either side of the chip. Standard commercial assembly techniques (chip placement with conductive epoxy and gold wirebonding) were used, providing a very economical and repeatable assembly.

5.6 Measured Results

This section describes the measured results of the fully integrated super-regenerative receiver.

5.6.1 Serial Interface

A three-wire SPI interface was used to facilitate on-chip biasing and frequency tuning while allowing a low pad count. This interface was fully functional, and allowed the the programming of eight 8-bit registers on the chip (2^{64} total states). These registers control on-chip biasing DACs, baseband transconductantances, and switched capacitor arrays. All receiver tuning and biasing was controlled via a laptop computer USB interface. The SPI interface utilizes a commonly-used, standard chip-to-chip protocol, allowing easy integration of this chip with other custom digital chips.

The SPI interface block integrates over 4000 high-V_T transistors and is composed of standard cell static $0.13\mu m$ CMOS logic. It is important to ensure that the leakage currents of such blocks are small relative to the rest of the receiver circuitry. Figure 5.11 shows the measured leakage current and power of this digital block over a range of supply voltages.

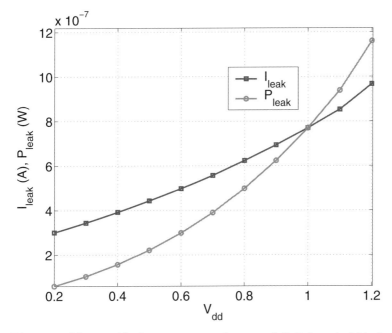

Fig. 5.11. Measured leakage current and power of digital control block

At the nominal supply voltage of 1V, the total leakage power of the SPI interface block is 770nW, which is negligible in relation to the other transceiver blocks. The total power consumption of the receiver when active is 380μW.

The linearity of the current source DAC was measured. Two important metrics for the accuracy of ADCs and DACs are the Integral Non-Linearity (INL) and the Differential Non-Linearity (DNL). The DNL is the incremental difference of each code width compared to the ideal code width (one LSB). A DNL less than +/- one LSB ensured there will be no missing codes. The INL is the integral of the DNL over the entire code range. See Figures 5.12 and 5.13 for the measured linearity performance of the current source DAC.

The DNL and INL errors are below +/- 0.5 LSB, indicating linearity accuracy that is sufficient up to 7 bits with no missing codes.

5.6.2 RF Front-End

This section documents the measured performance results of the radio frequency portion of the receiver. The BAW resonator capacitive tuning array operates as expected, and its performance matches theoretical calculations well. The measured frequency step size indicates a great deal about the health of the receiver; it reveals the accuracy of the switched capacitance, the loaded resonator Q, and the parasitic shunting capacitance of the resonator. The

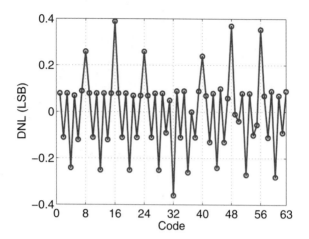

Fig. 5.12. DAC DNL Measurement

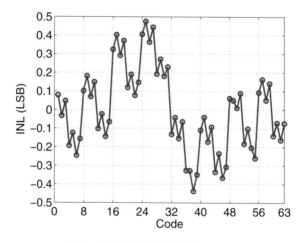

Fig. 5.13. DAC INL Measurement

frequency shift at each code was measured for three different chips and compared to the theoretical values as calculated by Equation 5.1. These results are plotted in Figure 5.14.

The measured values agree reasonably well with calculations. As Equation 5.1 predicts, the incremental frequency decrease is a function of the frequency code. As more capacitance is loaded on the tank, the frequency sensitivity to capacitance variation decreases. To achieve a linear frequency-code characteristic, the capacitors could be implemented as a thermometer-coded capacitor bank with increasing capacitance as a function of code.

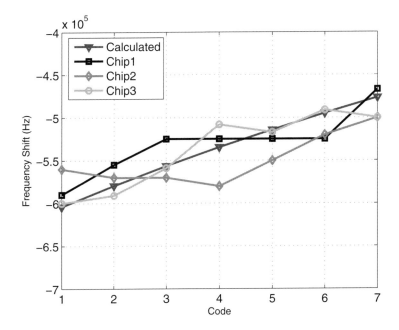

Fig. 5.14. Incremental frequency shift over all codes

A non binary-weighted scheme would result in a large increase in the required capacitance area.

The effects of random mismatch in the absolute and relative capacitive values are clearly visible. In addition, variation in the intrinsic resonator qualify factor will alter the total tuning range. Figure 5.15 shows the total measured frequency offsets for the three different chips. The matching of frequency tuning between the three CMOS/BAW systems is very close, allowing predictable frequency tuning. The maximum frequency variation that accumulated across the entire frequency range of the three chips was 72.6ppm, which is likely due to Q variation of the BAW resonators.

The super-regenerative oscillator amplitude will naturally vary across the oscillator tuning range. Figure 5.16 shows the measured variation in amplitude over the entire range.

Due to the relatively small tuning range and the high Q of the switched capacitance, the output amplitude variation across frequency code is small.

The phase noise of the detector oscillator was characterized over the frequency code range. See Figure 5.17 for the measured phase noise data.

The close-in phase noise is nearly identical to the other oscillators discussed in this book (for example: Figure 2.6 and Figure 2.12). Notice that the noise floor at large frequency offsets is higher than the other measurements presented here. Because there is no explicit RF output port, the phase noise

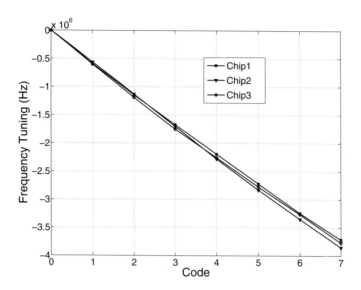

Fig. 5.15. Switched capacitor frequency tuning of receiver

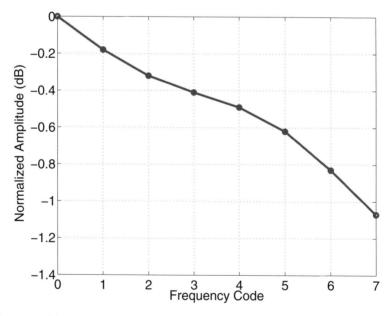

Fig. 5.16. Measured super-regenerative oscillator amplitude vs. frequency code

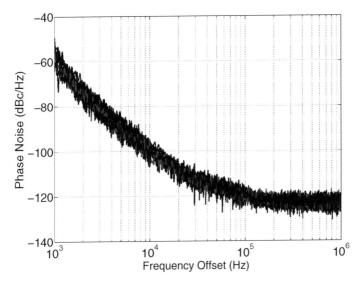

Fig. 5.17. Measured phase noise of detector oscillator over tuning range

was measured as a leakage signal at the receiver input port. This measurement technique results in a very low sinusoid amplitude, yielding a high measurement noise floor. Notice that the close-in phase noise actually improves slightly at the lower frequency range. This phenomenon is due to the increased RF signal power that results from the decreased oscillator tank impedance under the higher capacitive loading conditions.

The impedance match of the receiver was measured at nominal operating conditions and is shown in Figure 5.18. A comparison of input match on a test board and though direct die probing was made and plotted in Figure 5.18. As expected, the input match resonance decreases by approximately 200MHz when mounted on a test board. The quality of the input match as measured through die probing (approximately -10dB) is worse than expected due to a lower than expected source inductance.

The selectivity of the receiver was measured at a quench frequency of 100kHz. The selectivity measurement is important because it determines the interferer frequency offset and power levels that can be tolerated before the desired signal is corrupted. The measurement was performed by adjusting the input signal power for a demodulated BER of 1e-3 at various frequency offsets. See 5.19 for the corresponding receiver attenuation. The small signal receiver bandwidth is approximately 0.5MHz, and the upper and lower rejection is -47dB and -51dB, respectively. Super-regenerative receivers are typically very sensitive to exact bias levels. However, operation in the logarithmic mode relaxes the requirements on the absolute bias current accuracy. Figure 5.19 plots the selectivity for the nominal bias current of $400\mu A$. A higher current

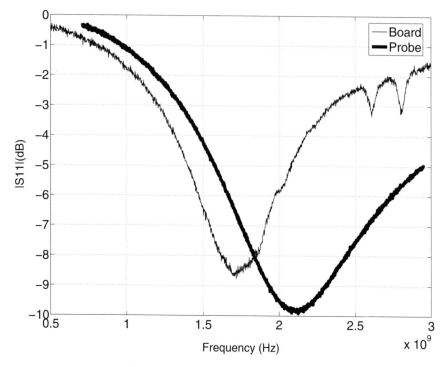

Fig. 5.18. Magnitude of receiver S_{11}

level of 455μA was also plotted for comparison. A 13% variation in the detector oscillator bias current results in minor variation in the receiver attenuation. Thus, the receiver bandwidth is relatively insensitive to variations in transistor biasing.

5.6.3 Temperature Compensation

As was shown in Figure 2.14, BAW resonators exhibit a temperature coefficient of approximately $-25\frac{ppm}{^\circ C}$. This temperature coefficient, though large, is well-behaved and easily compensated through switched capacitors. The integrated 3-bit capacitive array size was chosen to place the detector oscillator frequency within the -3dB bandwidth of the receiver (approximately 0.5MHz). This technique was experimentally verified with the super-regenerative receiver. See Figure 5.20 for the measured temperature data.

A temperature variation from 0 to 90°C would theoretically predict a frequency shift of 4.2MHz. However, by switching the capacitor array, this temperature drift can be compensated. The frequency deviation due to temperature was limited to the approximate RF bandwidth of the receiver. Temperature variation testing was performed with a *Temptronic LM01980* thermal

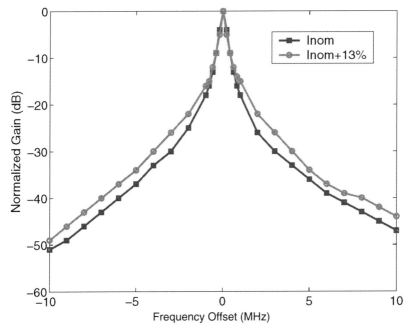

Fig. 5.19. Receiver selectivity profile for two bias current levels

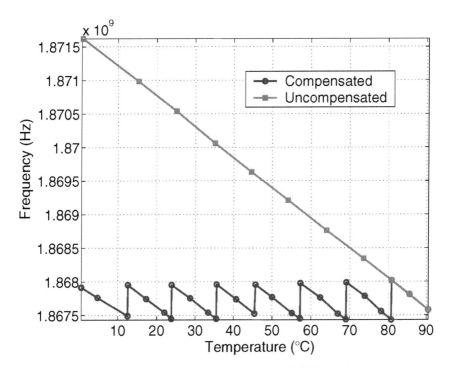

Fig. 5.20. Temperature compensation of BAW-based receiver

forcer. All temperature readings were verified with a *Fluke 52* thermometer with k-type thermal probe.

5.6.4 Baseband

A standalone pulse-width demodulator circuit was tested. The filter is programmable through the SPI interface. The measured performance of the filter across tuning range is shown in Table 5.1.

Table 5.1. Measured demodulator performance over tuning range

Parameter	Low Limit	High Limit
Cutoff Frequency (kHz)	14.5	45.0
Gain (dB)	6	16
Power (μW)	36	36

The filter worked as expected and exhibits good pole matching. Figure 5.21 shows the measured filter frequency response at the low and high frequency settings.

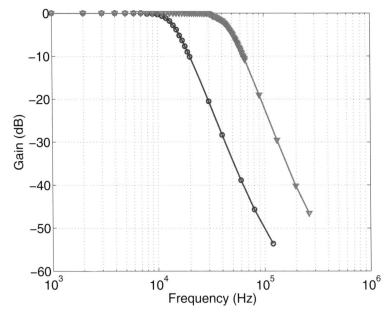

Fig. 5.21. Measured frequency response of baseband filter at the high and low bandwidth settings

Next, the filter was tested in-situ in its receiver environment. The replica biasing and non-linear filter worked as expected. This filter must provide baseband gain and reduce the quench tone ripple to below the receiver noise floor. The receiver quench signal was varied from 10kHz to the design value of 100kHz. Figure 5.22 shows the analog receiver output for three different quench frequencies.

As expected for the 10kHz quench tone pulse train, only a relatively large fundamental tone is visible. For a 100kHz quench tone, the ripple is reduced below the receiver noise floor, as desired.

5.6.5 Link Demonstration

To verify the link budget calculations and ascertain the robustness of the system, an end-to-end wireless link was implemented with two transceiver chips. A 10kbps, 27m indoor wireless link was demonstrated with a 400μW transmit power and a 380μW receiver power dissipation.

5.7 Discussion

This chapter has documented the design, implementation, and experimental results of a fully integrated 380μW super-regenerative receiver. The power breakdown of the receiver is given in Figure 5.23.

A majority of the power dissipation occurs in the RF front-end, where the required f_T of the transistors is approximately 2GHz. The required front-end current is approximately 250μW. The I/O consumes 15% of the receiver power budget. This is due to a conservatively-designed analog output buffer designed to drive a wide variety of test equipment loads. Likewise, conservative bias circuitry led to 9% of the chip power dissipation. Future modification of these components could lead to a near-negligible power dissipation contribution of these blocks.

Another option to reduce the interference levels and allow channelization is to use a CDMA approach. A CDMA scheme has been proposed for super-regenerative receivers that takes advantage of the excess receiver bandwidth [38]. This is yet another example of re-inventing the super-regenerative architecture by applying modern communication theory and circuit techniques.

The re-radiation of RF energy has been a traditional concern with super-regenerative receivers. This concern is especially valid relating to very dense

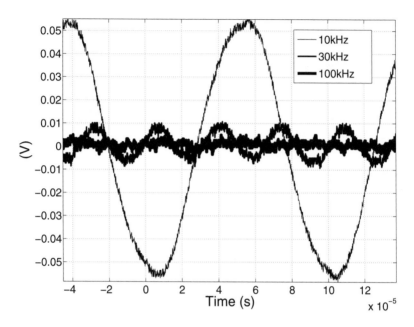

Fig. 5.22. Measured ripple of the pulse-width demodulator at quench frequencies of 10kHz, 30kHz, and 100kHz

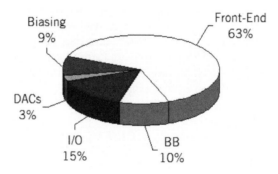

Fig. 5.23. Power breakdown of the fully integrated super-regenerative receiver

wireless sensor networks. The measured re-radiation of RF power of the receiver presented here is -69dBm, comparable to that of modern direct conversion receivers. At 2GHz, RF power drops off by approximately 35dB in the first meter. Thus, for node-to-node distances of greater than 1m, the incoming radiated receiver power from another node is below the noise floor.

6

INTEGRATION TECHNIQUES

One of the most aggressive goals of the wireless sensor research vision is the node volume specification. Achieving a 1cm^3 node volume (approximately the same volume as a 8-pin DIP op-amp package) is a tremendous challenging, and new integration techniques must be developed. This chapter describes technologies that significantly decrease the required implementation volume.

First, an IC/MEMS silicon reference clock is described. Ultimately, this component will obviate the need for a bulky, surface-mount quartz crystal resonator. It also enables system level degrees-of-freedom that are not currently available. For example, since customized frequencies can be defined lithographically, arrays of resonators at varying frequencies can be fabricated in a very small area on one substrate. Similar to the transition from discrete to integrated transistor circuit design, new design methodologies and architectures will accompany this new method of clock generation.

Secondly, flip-chip techniques are investigated, and two proof-of-concept prototypes are presented. Advanced CMOS/MEMS packaging and assembly techniques can improve the robustness, reduce interconnect parasitics, and decrease the form factor of these systems.

6.1 Silicon Reference Clocks

In wireless sensor network applications, a low frequency reference clock is needed for timing synchronization, synchronous digital circuit clocking, and A/D clocking. This clock is typically implemented by a CMOS oscillator tuned by a surface mount (SMT) quartz crystal. There are, however, serious drawbacks to this strategy. First, the surface mount (SMT) crystal (and accompanying SMT capacitors) often consumes a board area larger than the entire transceiver CMOS chip. This is unacceptable if the entire sensor node must be implemented in a 1cm^3 volume. Additionally, the SMT parts add to the bill-of-materials, increasing the cost of components and assembly. Furthermore, there is a finite selection of off-the-shelf quartz crystals, limiting the choice of operation frequency and motional resistance to that provided by commercially available crystals. Finally, the cost and size of quartz crystals greatly limits the number of frequency references available for use in the system. It would be desirable that the digital processor, the analog baseband, and the RF transceiver have the ability to use their own unique and optimized frequency reference. This chapter describes an effort to co-design a 16MHz MEMS resonator[1] with custom designed $0.13\mu m$ CMOS oscillators. This technology merge would allow the fabrication of very small (approximately $10000\mu m^2$), fully integrated MEMS resonators on top of standard CMOS circuitry.

Two oscillators were designed: one optimized for low phase noise and the other optimized for low power consumption. Both include an amplitude control loop to provide a stable output amplitude at a user specified value. The low power oscillator[2] will not be discussed here. This chapter will describe the design, implementation, and testing of a $100\mu W$, low phase noise reference clock for wireless sensor networks.

6.1.1 Silicon Resonator Background

Post-processing of MEMS structures on top of CMOS using low temperature processing is possible using SiGe structural layers and Ge sacrificial layers [39]. This process has been further modified to fabricate very small electrostatic gaps. Two custom 16MHz resonators were designed and fabricated in the Berkeley Microlab Blade process [40]. This is an experimental MEMS process developed at U.C. Berkeley to allow the fabrication of narrow, high aspect-ratio gaps for efficient capacitive transduction. The process uses germanium "blades" that are defined and etched as thin as 50nm. The silicon-germanium (SiGe) structural layer is then deposited over the blades. When the sacrificial Ge blades are removed, a 50nm gap remains. A brief process outline is shown below:

- **Mask 1 - Substrate Contact:** The contact substrate to the n+ wafer is patterned and etched through a SiO_2 and Si_3N_4 barrier layer.
- **Mask 2 - Interconnect:** The $1\mu m$ Al/TiN interconnect layer is deposited, patterned, and etched. The contact substrate to the n+ wafer is patterned and etched through the SiO_2 and Si_3N_4. A $1\mu m$ resolution is possible.
- **Encapsulation:** The interconnect is encapsulated and planarized by a SiO_2 deposition/CMP step.
- **Ge Deposition:** A $2.5\mu m$ thick Ge layer is deposited (350°) along with a $0.5\mu m$ SiO_2 hardmask.

---

[1] Designed by Emmanuel Quévy of the Berkeley Sensor and Actuator Center (BSAC), U.C. Berkeley

[2] Designed by Nathan M. Pletcher, EECS Dept., U.C. Berkeley

- **Mask 3 - Blade Definition:** The SiO_2 hardmask is patterned and etched, allowing etching of the Ge blade layer. This results in 50nm thick, $2\mu m$ tall Ge blades.
- **Mask 4 - Interconnect Contact:** Contacts are patterned and etched through the Ge sacrificial layer down to the Al/TiN interconnect layer.
- **SiGe Structural Definition:** The SiGe structural layer is deposited ($425°$) and planarized with CMP.
- **Mask 5 - Pad Formation:** Aluminum pads are deposited, patterned, and etched on the SiGe structural layer.
- **Ge Release:** The Ge is etched away, releasing the SiGe structural layer with 50nm gaps.

The use of a Ge sacrificial layer and a SiGe structural layer allows a high selectivity in the release process. To limit the SiGe gap increase to 10nm, a 40 minute release step is possible, which allows a $20\mu m$ lateral etch of the Ge sacrificial layer. A simplified process sequence is shown in Figure 6.1 [40].

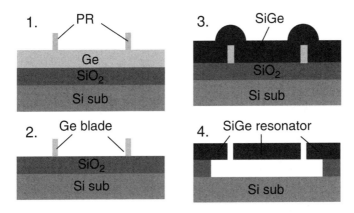

Fig. 6.1. Cross-sectional process sequence. Reprinted with permission of IEEE

This process was used to fabricate two different 16MHz resonators. One is composed of wineglass mode rings. The other consists of Lamé mode plates. See Figure 6.2 (a) and (b) for the Ansys simulation of the two resonators.

From an electrical point of view, the series motional resistance (R_x) of the resonator is one of the most important parameters. A low R_x is desirable as this allows a low series impedance and a high parallel impedance, facilitating the design of oscillators and tuned amplifiers. For an electrostatic resonator, the motional resistance is given by Equation 6.1.

$$R_x = \frac{kg^4}{\epsilon^2 A^2 \omega^2 Q V_{bias}^2} \tag{6.1}$$

   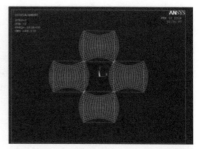

(a) Coupled wineglass mode resontor   (b) Coupled lame mode resonator

Fig. 6.2. Simulation of the coupled wineglass and lamé mode resonators

Where g is the physical electrostatic gap, A is the electrode area, ω is the natural frequency, and V_{bias} is the applied DC bias voltage. It is apparent that reducing the physical gap and increasing the DC bias voltage are the most effective ways of decreasing the resonator motional resistance. Capacitively driven and sensed resonators require a DC bias voltage to increase their transduction efficiency. Varying the bias voltage allows tuning of the series resonant frequency. For a nominal bias voltage of 3.3V, the targeted resonator parameters for the wineglass mode resonator are given in Table 6.1. Due to the experimental nature of the Blade process, precise control over the

Table 6.1. Predicted resonator parameters for different achieved gaps

Parameter	50nm	100nm
C_o	1fF	1fF
L_x	0.20	3.22
C_x	482aF	30.1aF
R_x	2.04kΩ	3.27kΩ

sub-lithographic gaps is not possible in the first revision. The expected fabricated gap sizes are between 50 and 100nm. Table 6.1 provides the expected resonator parameters for both ends of the anticipated tolerance range. The most important implication for the oscillator design is that the circuit must tolerate a motional resistance from 2kΩ to 3.5kΩ with an expected resonator Q of approximately 10,000.

6.1.2 Circuit Analysis and Design

To minimize the phase noise of the reference oscillator, the sinusoidal power exciting the resonator should be maximized. Ultimately, this power is limited by the power dissipation budget of the electronics, the achievable voltage swing of the process used, and the linearity of the MEMS resonator transducer. To this end, the series resonant mode of the resonator was used for the low phase noise version. The series resonance presents a lower impedance to the oscillator, allowing a higher oscillator loop power for a given supply voltage. Additionally, capacitively driven and sensed resonators typically exhibit high motional resistances ($>1\text{k}\Omega$) and extremely high ($>1\text{M}\Omega$) parallel resistances. To avoid detuning the resonator excessively, it is easier to drive the resonator in its series mode [41]. However, driving a resonator at its series resonance requires an entirely different sustaining amplifier compared to that used for a parallel resonant oscillator. See Figure 6.3 for a conceptual oscillator diagram.

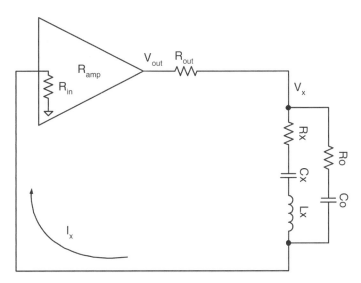

Fig. 6.3. Conceptual schematic of 16MHz clock oscillator

In order to avoid degrading the quality factor of the mechanical resonator when driven in its series resonant mode, the resonator should be driven and sensed with a low impedance. Thus, the sustaining amplifier must present a transimpedance to the resonator. This amplifier drives the resonator with a voltage and senses the current passing through it, providing positive feedback at one precise frequency. In Figure 6.3, the amplifier sustains oscillation with a transimpedance gain R_{amp}, and exhibits a finite input and output resistance. Equation 6.2 shows the Q degradation that occurs with a finite transimpedance amplifier input/output resistance.

$$Q_{loaded} = Q_{unloaded}\frac{R_x}{R_x + R_{in} + R_{out}} \qquad (6.2)$$

It can be shown that the oscillator loop gain at the series resonance of the mechanical resonator is given by Equation 6.3.

$$A_L = \frac{R_{amp}}{R_x + R_{in} + R_{out}} \tag{6.3}$$

Thus, if the acceptable Q degradation is 50%, a transimpedance gain of $2R_x$ is needed. To accommodate resonator variation in Q and R_x, adjustable gain, R_{in}, and R_{out} is desirable. In addition, for stable and efficient oscillation, an amplitude control loop (ACL) is necessary.

A new oscillator circuit topology was designed to meet these requirements. See Figure 6.4 for the oscillator schematic. The oscillator consists of a tran-

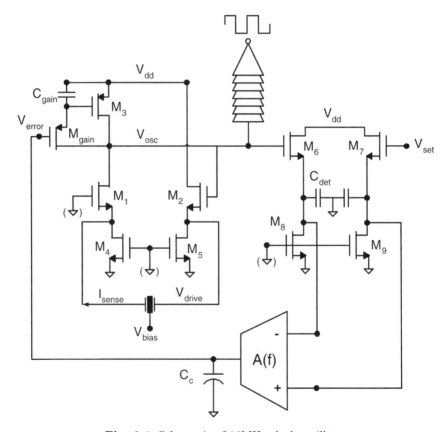

Fig. 6.4. Schematic of 16MHz clock oscillator

simpedance amplifier with adjustable gain, level detection circuitry, and a compensated OTA to close the amplitude control loop. The transimpedance amplifier comprises transistors M_1-M_5. The current from the mechanical resonator is sensed by the low impedance of the source of transistor M_1. The signal then undergoes amplification by a factor of $g_{m1}R_{gain}$, where R_{gain} is the effective resistance of M_{gain} in the linear operating regime. The structure formed by M_{gain}, C_{gain}, and M_3 provides an effective inductance given by Equation 6.4.

$$L_{eff} = \frac{R_{gain}C_{gain}}{g_{m3}} \tag{6.4}$$

It can also be shown that a flat load impedance (amplifier gain) of R_{gain} is achieved at frequencies over

$$f_{passband} = \frac{g_{m3}}{2\pi C_{gain}}Hz \tag{6.5}$$

Thus, for frequencies higher than $f_{passband}$, the gain of the amplifier can be controlled by adjusting the gate voltage of M_{gain}. At very large values of R_{gain}, the gain is limited by a shunting resistor (not shown). Through this method, the maximum amplifier gain is limited to approximately 100kΩ. In this regime, maximum gain is reached at even lower frequencies. This amplified signal is buffered by the M_2, M_5 follower. Thus, the input and output impedances of the transimpedance amplifier are $\frac{1}{g_{m1}}$ and $\frac{1}{g_{m2}}$, respectively.

The proposed oscillator topology allows independent control over the input impedance, output impedance, and amplifier gain. The minimum anticipated resonator R_x was 2kΩ and the allowable Q degradation factor was set at 2. Thus, $R_{in} + R_{out} < 2k\Omega$. For moderate inversion operation of M_1 and M_2, yielding $\frac{g_m}{I_d} = 20$, the required bias current in each leg is 50μA. The gain of the transimpedance amplifier is shown in Figure 6.5. For $\frac{g_{m3}}{I_d} = 10$, and $C_{gain} =$

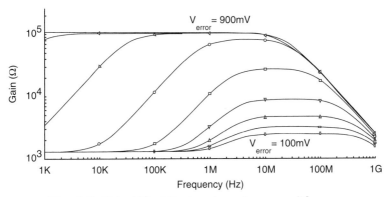

Fig. 6.5. Gain (Ω) of the transimpedance amplifier

20pF, Equation 6.5 predicts the amplifier passband beginning at 4MHz for low gain values. This is evident in Figure 6.5. In addition, the passband clearly expands as the gain is limited by shunting conductance. The valid error voltage range, 100mV to 900mV, is determined by the output swing capability of the amplitude control loop OTA. Over this voltage range, the linear resistance of the PMOS transistor M_{gain} increases. Thus, the amplifier gain varies from 2kΩ to 100kΩ, covering all possible ranges of resonator motional resistances. For the nominal value of $R_{amp} = 4$kΩ, the required V_{error} is 300mV.

The amplitude control loop consists of a signal level detector (M_{6-9}), an OTA, an external signal V_{set}, and the resulting error signal V_{error}. See Figure 6.6 for the OTA schematic. A current mirrored OTA with a common

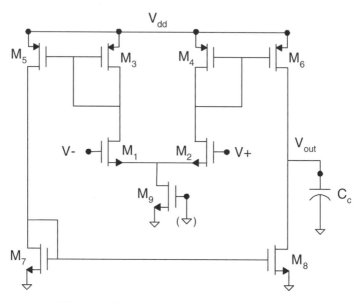

Fig. 6.6. Schematic of amplitude control OTA

source, non-cascoded output was the chosen OTA topology. This choice is appropriate because the loop accuracy requirements are not stringent; high OTA gain is not necessary. The non-cascoded output stage allows a large output voltage range, which is necessary to allow a large variation in the variable gain of the transimpedance amplifier. The input transconductors, M_{1-2}, are large ($\frac{50\mu m}{0.5\mu m}$) and laid out in a common centroid configuration to minimize the offset voltage of the OTA. The open-loop DC gain of the OTA is approximately 50dB, and the circuit consumes 2μA from a 1V supply.

Although not shown in Figure 6.4, the oscillator loop is AC-coupled to the amplitude detector. The coupling circuitry has a high pass characteristic with a 1.6MHz cutoff frequency. AC-coupling allows the removal of the oscillator DC offset and pulls both detector inputs to the same default voltage. Thus, the loop defaults to a zero oscillation amplitude. To begin oscillation, V_{set} is increased above its equilibrium voltage by the desired oscillation amplitude. This increases the positive OTA input, reducing the V_{gs} of transistor M_{gain}, increasing the transimpedance amplifier gain. Thus, the loop gain of the oscillator increases. As oscillation ensues, the negative OTA terminal is driven to the same value as the positive terminal. See Figure 6.7 for the transient simulation of the oscillator amplitude control loop convergence process. Due

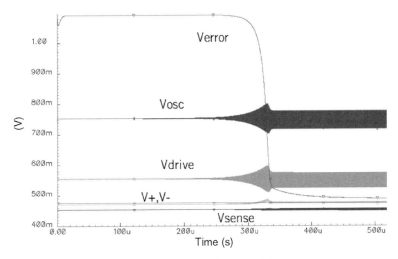

Fig. 6.7. Simulation of amplitude control loop convergence

to the initial set voltage on V+, the error signal V_{error} is driven high by the OTA. The transimpedance amplifier gain is set to its maximum value, allowing a high initial loop gain. At the onset of oscillation, V- increases to meet V+. Meanwhile, V_{error} settles to its equilibrium value with no visible ringing. Notice the relative amplitudes of the three oscillator signals V_{osc}, V_{drive}, and V_{sense}. The simulated amplitudes correspond well to calculated levels using the model in Figure 6.3. The dominant pole of the amplitude control loop is set by the resonator time constant $\frac{2Q}{\omega}$. Thus, as the quality factor of the resonator decreases, the loop stability degrades. The amplitude control loop is stable down to a resonator Q of 1000. Below this value, the response time of the oscillator is fast enough to compromise the stability of the loop. The oscillator drives a seven-stage inverter chain, which was optimally sized to drive a 10pF test equipment load.

See Figure 6.8 for the CAD layout of the complete oscillator. Shown in the layout plot is a 500μm x 500μm section of the die that contains the active circuitry and the resonator flip-chip pads. The completed resonator will be flipped and bonded directly to this inner padring to avoid board parasitics. There are two drive pads, two sense pads, and two sets of ground pads to accommodate multiple mechanical resonant structures. The resonator bias pads are also shown. Although only one bias contact is necessary, two pads allow current flow through the resonator, enabling temperature compensation at the

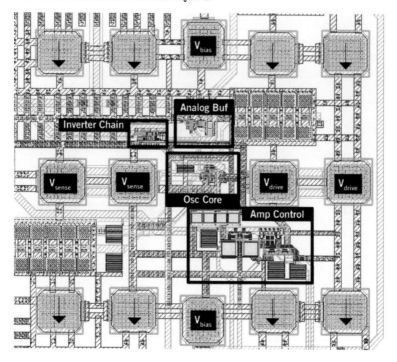

Fig. 6.8. Layout of 16MHz reference clock

expense of dissipated power. The oscillator core is situated directly between the drive and sense pads. Visible in the amplitude control loop are the coupling capacitors, the amplitude detector, and the OTA. An analog buffer with 50Ω drive capability is included for diagnostic purposes. The oscillator consumes approximately 100μW, and the amplitude control circuitry consumes approximately 3μW.

6.2 Flip-Chip Packaging

Packaging and assembly of the transceivers presented in Chapters 3, 4, and 5 has consisted of COB bonding on a test board. The connection of the BAW resonator to the CMOS circuitry has been accomplished through direct wire-bonding of one chip to the other, as shown in Figure 4.7. Although this assembly method provides an inexpensive, sufficiently low parasitic interconnect of the resonator to the COB, more advanced packaging techniques are desirable. For example, to minimize bondwire length, the two chips must be placed in close proximity. Thus, a very sparse pad structure must be present on the perimeter of the CMOS chip. For pad-limited implementations, this requirement translates into an increased CMOS die area, which is prohibitively expensive. In addition, as resonator technology progresses and multiple frequencies become available, even more die perimeter would be needed to realize the CMOS/resonator interconnect. Thus, it would be preferable that the resonators be flipped directly on top of the CMOS chip, allowing arbitrary placement on the chip. This methodology eliminates the need for pre-allocating space on the chip perimeter to accommodate resonator connections. In addition, the interconnect inductance would be smaller, reducing the risk of parasitic modes in oscillator or amplifier operation. To test this strategy, two prototypes were implemented. Both prototypes utilized a μ-capped (micro-capped) PCM test structure from Avago Technologies [42]. This technology allows wafer-scale hermetic packaging of BAW resonators. Thus, epoxy can be flowed under the inverted resonator without risk of damaging the delicate Aluminum Nitride (AlN) membrane. This method uses one conductive and one non-conductive epoxy for the interconnect and mechanical support, respectively. See Table 6.2 for the relevant epoxy characteristics.

Table 6.2. Flip-chip epoxy characteristics

	Conductive	Insulating
Epoxy	H20E-PFC	U300
Resistivity $(\Omega - cm)$	$4x10^{-4}$	$1x10^{14}$
Relative permittivity	-	4.1
Dielectric breakdown	-	$17.71 \frac{V}{\mu m}$

The electrical characteristics of the epoxy determine the parasitic losses and coupling that occur when the resonator is flipped onto the CMOS chip.

6.2.1 Oscillator Flip-Chip

As a proof-of-concept, an encapsulated FBAR was flipped onto a well characterized $0.18\mu m$ CMOS oscillator presented in Section 2.3. Due to pad limitations, it was necessary to cantilever the resonator off the edge of the CMOS chip. Although this presented obvious complications for the flip-chip procedure, success with this effort would bode well for the reliable assembly of more conservative mounting orientations. The assembly process consisted of the die-level bumping of the CMOS chip with gold bondwire. Next, the FBAR chip was mechanically bonded to the bumps and non-conductive epoxy was underfilled between the chips. See the photograph of the completed flip-chip in Figure 6.9. The technique yielded working oscillators in a very small form-factor. Direct comparison the wirebonded version was not possible since different resonator technologies were used.

Fig. 6.9. BAW flipped onto a 0.18μm CMOS oscillator

6.2.2 Super-Regenerative Flip-Chip

To demonstrate a fully integrated, 1mm^2 receiver, a BAW resonator was flipped onto the super-regenerative receiver discussed in Chapter 4. Figure 6.10 (a) shows an SEM image of the flip-chipped system. A detailed view

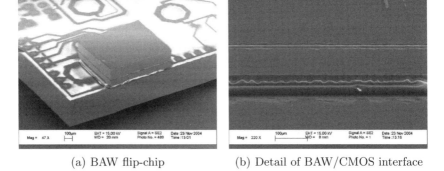

(a) BAW flip-chip   (b) Detail of BAW/CMOS interface

Fig. 6.10. SEM image of super-regenerative flip-chip

of the interface is shown in Figure 6.10 (b). The die-to-die distance is approximately 60μm, which is filled with *Epo-tek U-300* non-conductive epoxy.

The total thickness of the receiver is increased by a factor of less than two. The entire CMOS/BAW assembly could easily be packaged in a single, standalone unit. Aggressive use of this technique would allow placement of BAW resonators over the entire surface of the CMOS transceiver, including over the on-chip planar inductors that typically occupy much the transceiver area. One concern is that the flip-chipped BAW die would couple to the planar inductors, decreasing the self-resonant frequencies and reducing the quality factor. To quantify the effect of this phenomenon, the input matching (S_{11}) of the super-regenerative receiver was probed with and without a BAW resonator flip-chipped over the input matching inductor. See Figure 6.11 for the coverage of the planar inductor by the BAW resonator. Approximately 40% of the 10nH inductor is directly covered by the resonator. With a typical die-to-die distance

Fig. 6.11. Coverage of planar inductor by BAW resonator

of 60μm and a *U-300* epoxy dielectric constant of 4.1, the expected coupling between the inductor and resonator is weak. Figure 6.12 quantifies the effects of the flip-chip assembly on the input match. The receiver front-end was biased to its nominal conditions. The input match was measured with a *Cascade ACP* GSG probe and an *HP 8719* Network Analyzer. In the implementation where the BAW resonator is wirebonded to the CMOS die, the S_{11} peak is -27.5dB at 1.894GHz. For the receiver with the flip-chipped resonator, the S_{11} peak is -27.8dB at 1.891GHz. The input match frequency shift of 0.16% is negligible, indicating little loss in performance if planar inductors are covered by a flip-chipped resonator. These results are summarized in Table 6.3.

The direct flip-chipping of BAW resonators onto CMOS die provides a high performance, robust, easy to package option for the implementation of fully integrated transceivers.

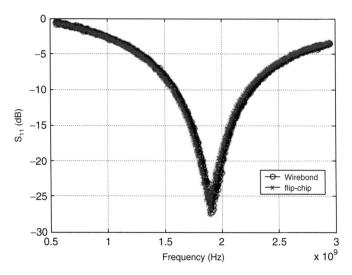

Fig. 6.12. Super-regenerative S_{11}: Effect of flip-chip proximity on planar inductors

Table 6.3. Input matching performance of flip-chip vs. wirebonded die

Assembly	Measured S_{11} peak	Frequency
wirebonded	-27.5dB	1.894GHz
flip-chip	-27.8dB	1.891GHz

6.3 Conclusions

New applications like peer-to-peer wireless sensor nodes and implantable electronics demand extremely small and inexpensive electronics. These systems demand a paradigm shift in packaging techniques. Multi-layer printed circuit boards with surface-mount components are too bulky for these applications. Components like quartz crystal resonators are much larger than the actual CMOS chips. This chapter described the design of a reference clock using SiGe MEMS resonator to replace quartz crystals. In addition, advanced packaging techniques to directly bond BAW resonators to the CMOS chips were investigated.

7

ULTRA-LOW POWER RADIO IN A PACKAGE USING ULTRA-WIDE BAND TECHNOLOGY

By Julien Ryckaert and Steven Sanders. IMEC, Leuven, Belgium.

7.1 Introduction

Wireless sensors have to demonstrate sufficient autonomy in order to make the dream of a seamless ubiquitous environment come through. This autonomy requires the sensor nodes to find means to capture their energy from the ambient through scavenging techniques. However, the limited power delivered by these energy sources sets an arduous limit on the order of a $100\mu W$ [43] on the operating power of the radio. Only new approaches in the design of the radio can break the energy gap with actual radio implementations. Different air interfaces based on traditional approaches have been proposed for the communication of sensor nodes at low power. However, none of them has proven the ability to reduce the power consumption by the required orders of magnitude while offering the necessary communication performances [44]. Moreover, the overall size of the packaged radio should not exceed the required form factor of a sensor node limiting the complexity of the system, the size of the energy storage and the type of scavenging technology. Recently, a novel air interface based on communication using wideband signals has attracted a large attention from the wireless community, the so-called Ultra-Wide Band (UWB) communication [45]. This happened mainly due to a decision of the FCC [46] to release a large UWB spectral mask. This type of communication initially generated a lot of interest for short range high data rate communications [47], while deviating from the original pulsed signal format of traditional UWB communication. On the other hand, and more recently, the low power radio community gained an increasing interest in UWB technology for low-power low data-rate sensor network communication [48]. Indeed, for example, in its pulse-based format, the short time duration of the useful signal allows turning off the radio between the transmitted and received impulses offering a large potential in the reduction of the radio static power consumption. Also, thanks to the high Ft of todays silicon transistor devices, CMOS circuits can operate as wideband circuits at RF frequencies offering the potential for highly integrated UWB systems in a single-chip. This chapter analyzes and demonstrates through examples of implementations the potential of impulse radio UWB for low power sensor network communication. Section II describes the UWB air interface envisioned for sensor network communication and shows the ability of UWB technology to reduce both the complexity as well as the power consumption of a sensor radio. Section III then describes some circuit implementations of a UWB radio. Finally section IV addresses at a higher level the important packaging aspects of a heterogeneous system such as a sensor.

7.2 UWB for Sensor Networks

7.2.1 Air Interface Definition

The Federal Communications Commission (FCC) has authorized UWB communications between 3.1GHz and 10.6GHz. Although the regulations on UWB radiation define a power spectral density (PSD) limit of -41dBm/MHz, there are very few regulations on the definition of the time-domain waveform. The latter can then be tailored for low hardware complexity as well as low system power consumption. In pulse-based UWB, the transmitter only needs to operate during the pulse transmission, producing a strong duty cycle on the radio and the expensive baseline power consumption is minimized. Moreover, since most of the complexity of UWB communication is in the receiver, it allows the realization of an ultra-low power, very simple transmitter and shift the complexity as much as possible to the receiver in the master. However, the impact of the type of UWB signal chosen on the communication performance and on the complexity of the radio implementation must be carefully analyzed. The minimum bandwidth of a UWB signal is usually 500 MHz. Indeed, various UWB standard proposals ([48], [47]) have subdivided the entire UWB spectrum in 500MHz sub-bands as a solution to mitigate against strong interferers, to improve the multiple access and to compose with the different regulations on UWB emissions worldwide. Therefore, in order to comply with these regulations and standards, the generated pulses of UWB impulse-radio (UWB-IR) approaches must fulfill stringent spectral masks that can feature such low bandwidths. This poses a serious challenge for the pulse generation of UWB-IR transmitters. The gaussian pulse shape and its derivatives are often assumed in the literature as typical UWB transmitted impulse waveforms [49]. These waveforms suffer from several major drawbacks. First, the spectrum must be specified through both the original gaussian standard deviation parameter and the derivative order. The calibration of the parameters defining the gaussian pulse is a difficult task. For example, on Figure 7.1-(a), a variation of -10% (case B) and -20% (case C) on the 9th order derivative of the gaussian pulse (case A) causes the center frequency to drift significantly. Moreover, the distortion is asymmetric in linear frequency with the higher frequencies being more affected than the lower frequencies. Second, the generation of gaussian shape pulses is not obvious using standard CMOS circuits. Traditional impulse generators generate wideband monocycles that approximate gaussian shapes using very specific components such as step recovery diodes [50] or choke inductors [51]. These techniques suffer from a very limited tunability and require complex packaging techniques, thereby substantially increasing the system cost. Finally, as the waveform must comply with narrow spectral masks, the order of the Gaussian derivative pulses of traditional approaches have to be absurdly high in order to comply with the stringent UWB spectral masks [49]. Another way to realize short high frequency UWB signals is by

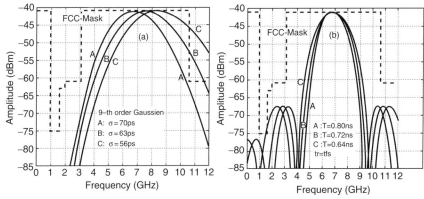

Fig. 7.1. Impact of parameter variations on the PSD (a) of the gaussian pulse, (b) on the triangular pulse

gating an oscillator as was proposed in [52]. The oscillator center frequency is then defined independently from the bandwidth by the gate duration. This class of UWB systems are defined here as "carrier-based UWB" impulse radio due to the presence of a "weak" carrier inside the pulse. However, since a direct gating of the center frequency features a rectangular shape, the high sidelobe power must be filtered. Since variable RF channel select filter is not an option, the shape must be filtered before translated to an RF signal. In low-cost and low power applications, this filtering operation should require circuitry with minimal complexity and preferably compatible with full-CMOS integration. Therefore, the triangular signal pulse appears to be a perfect compromise. The smooth shape of the triangular waveform provides a sidelobe rejection of more than 20dB with most of the power confined in the useful bandwidth. The triangular signal is relatively easy to generate in standard CMOS circuits by for instance charging and discharging a linear capacitor. Figure 7.1 -(b) shows, as a comparison to the gaussian pulse, a variation of -10% (case B) and -20% (case C) on the duration of a triangular waveform (case A). The baseband shape being independent from the carrier frequency, these variations can easily be compensated through some calibration loops.

7.2.2 UWB System Implementation

In order to create shaped carrier-based pulses, an RF oscillating signal can be multiplied with a specific baseband shape. This operation, shown on the block diagram of Figure 7.2, requires however the spectrum to be perfectly calibrated. Two control loops are therefore required, one fixing the bandwidth and the other the center frequency. The UWB pulses can be modulated either in time, featuring Pulse Position Modulation (PPM), or in phase, for Binary Phase Shift Keying (BPSK). PPM modulation is achieved by adding a predefined variable delay on the triggering signal of the baseband shape generator. For BPSK, the carrier can simply be inverted to produce the antipodal symbol. Since the power consumption has to be minimized, the oscillator producing

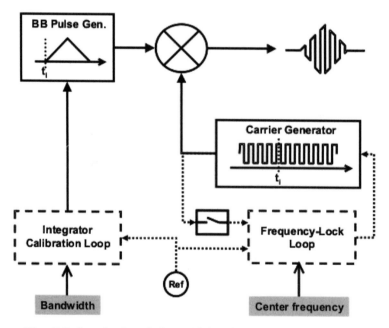

Fig. 7.2. Impulse-based ultra-wideband transmitter architecture

the carrier could be switched off after each pulse transmission and reactivated when a pulse must be transmitted. This duty-cycling can be done on the oscillator but not on its control loop since frequency-locking is a typically slow process. Therefore this operation imposes the oscillator to operate in free-running mode during the transmission of a pulse and the center frequency must remain stable over the oscillator switching process during the pulse sequence. Thanks to the wideband pulses, the requirement on the center frequency drift is less stringent than for narrowband applications. In the latter, a drift on the center frequency is usually not acceptable since the ratio between signal bandwidth and center frequency is very small. In UWB, a drift of the signal spectrum by a fraction of the center frequency still keeps most of the power inside the reception window. Let us consider a receiver using an oscillator at frequency $f_c + \Delta f$, while the transmitter uses the central frequency f_c. We consider a quadrature receiver. Without loss of generality, we assume that the receiver has a branch exactly in-phase with the received signal in the center of the pulse, and the other (in I/Q context) orthogonal. With a symmetric pulse, it means that all the received power of the detector will come from that branch, and none from the other as it will lead to opposite values in the first and second half of the pulse. Assuming that by reaching one extremity of the pulse, the useful branch has shifted by 90° compared to the received signal, this would shape the correlation over the pulse duration by a sine function between 0 and π, leading to an average correlation of $2/\pi$ (normalized integral of the sine function) for a rectangular pulse, instead of a correlation of 1 with perfect phase alignment. This is a loss of almost 4 dB. However, with a more realistic pulse shape (such as a triangle), the loss will be reduced as the components close to the extremity where the reduction is highest have a reduced weight. The triangle shape gives $8/\pi^2$, or 1.8 dB. Hence, this phase shift of 90° over a half-pulse is used as worst-case acceptable drift on the frequency offset. Practically, much less will be observed. Indeed, the frequency drift can be computed from the 90° phase shift spec:

$$2\pi \Delta f \frac{T}{2} = \frac{\pi}{2} \tag{7.1}$$

using the pulse duration T and the frequency offset Δf. The corresponding relative tolerance on center frequency can be computed based on (7.1):

$$\alpha = \frac{\Delta f}{f_c} = \frac{1}{2T f_c} \tag{7.2}$$

Due to the FCC regulations, the minimal bandwidth at -10 dB is 500 MHz. This enables the computation of the pulse duration from the time-bandwidth product, which is for a triangular pulse with -10 dB bandwidth equal to 2.23. This leads to T, equal to 2.23/500MHz. Since UWB communication operates between 3.1GHz and 10.6GHz, and the signal must have a bandwidth of at least 500MHz. The worst-case situation will be 500MHz bandwidth around 10GHz (5% relative bandwidth), allowing 1.8-dB degradation to occur for a center frequency drift of $\alpha = 1/(2.23/500MHz \times 2 \times 10GHz) = 1\%$. As an example, for a pulse with a center frequency $f_c = 4GHz$ we have a maximum variability of $\Delta f = 40MHz$ over the burst duration. Typical burst duration is around few ms. The center frequency stability of free-running oscillators is orders of magnitude lower, since they are mainly due to rather slow processes such as temperature and supply variations. Hence, the oscillator can be kept in free-running mode for the burst duration. For the bandwidth, the triangular-shape signal can be created with well-controlled parameters using a simple integrator circuit. A circuit principle is shown in Figure 7.3. The rising and falling slopes of the triangular waveform depend then on the current to capacitor ratio of the integrating circuit. In order to define a precise

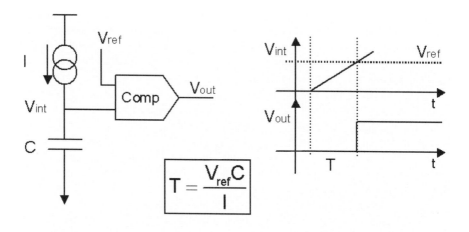

Fig. 7.3. Integrator circuit to generate the ramping slope of the triangular signal

pulse bandwidth, this ratio must be accurately calibrated. Therefore, the integration period of time $\frac{V_{ref}C}{I}$ must be compared to the time reference of the system with a simple delay locked-loop (DLL).

7.3 UWB Radio Design

The UWB concepts developed in section 7.2 have been demonstrated in circuit implementations and are described in this section. First, an ultra-low power pulse generator [53] where all circuits are off between each pulse transmission is demonstrated. It offers a large reduction in the transmitter static power consumption while providing the necessary flexibility in the pulse spectrum. In the second part of the section, a low-power UWB receiver for impulse radio is presented [54]. It contains the radio front-end as well as analog baseband

functionnality including a high speed analog to digital converter on a single
0.18μm CMOS chip.

7.3.1 UWB Pulse Generator Design

The pulse generator architecture is presented in Figure 7.4. A triangular pulse
generator and a ring oscillator are activated simultaneously. The triangular
signal is multiplied with the carrier created by the oscillator, resulting in an
up-converted triangular pulse at the output. The triangular waveform has a
duration that can be adapted in accordance with the desired bandwidth. A
gating circuit (ring activation circuit in Figure 7.4) activates the ring oscillator when a pulse must be transmitted, avoiding useless power consumption
between the pulses. This motivates the choice of a ring type of oscillator since
its startup time is very low while the high phase noise problem is obviously not
an important issue in the generation of wideband signals. In this subsection,
we describe each of the building blocks of the pulse generator.

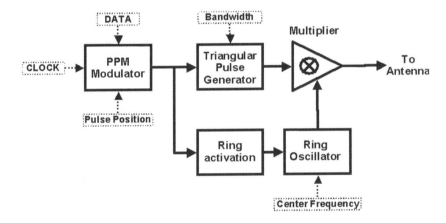

Fig. 7.4. Block diagram of the pulse generator

Precise Timing Generation

The difficulty in UWB communication is the precise realization of very short
time delays. The circuit shown in Figure 7.3 uses a charging current flowing
into a capacitor and compares the resulting voltage to a reference voltage
generating a precise timing. This timing depends on the capacitor value, the
integrating current and the threshold voltage. Therefore these quantities have
to be calibrated with a reference circuit.

The PPM Modulator

In Figure 7.4, a PPM modulator is placed just after the triggering signal. Based on the current integrator described above different timings, each corresponding to a pulse position, are generated by switching between capacitors depending on the value set at the data input. A precise time difference can be defined by tuning the capacitor value. This operation is done via the "pulse-position" input on Figure 7.4. The circuit diagram is shown for two capacitor values in Figure 7.5. The time difference between two pulse positions is given

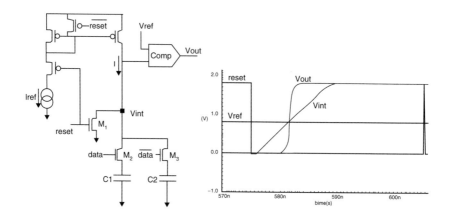

Fig. 7.5. PPM modulator circuit

by the difference in load capacitance seen at the integration node. That is:

$$\Delta T = \frac{V_{ref}}{I}(C_1 - C_2) \tag{7.3}$$

For example, with a reference voltage V_{ref} of 1V and capacitor values of 300fF and 150fF for C_1 and C_2, a PPM delay of 1ns is obtained for a current of 150μA. The circuit operates as follows. Once the reset signal sets to low (falling edge), the PMOS current source is switched on and the integration node, released by transistor M1, can charge either C_1 or C_2 with a current I such that V_{int} increases. Once V_{int} reaches the reference voltage, the comparator output V_{out} switches to a logical one. The voltage V_{int} however continues to increase hereby bringing the current source in triode region. At the next rising edge of the reset signal, transistor M1 shorts the integration node V_{int} back to ground, the PMOS current source is switched off and the circuit is ready for a next integration. A simulation example of the different voltages is shown in Figure 7.5. This technique can obviously be generalized to more than two delays by adding more capacitors.

The Pulse Shaping Circuit

The pulse shaping circuit is realized with a similar capacitor charging principle. If the current is constant and the capacitor is linear, the capacitor voltage increases linearly. If the current is inverted during the integration, the capacitor is discharged linearly and a triangular voltage is created. Varying the current and the capacitor modifies the charging and discharging slopes and the duration of the triangle can be tuned. This operation is realized with the circuit in Figure 7.6. The charging operation is similar to that of the PPM

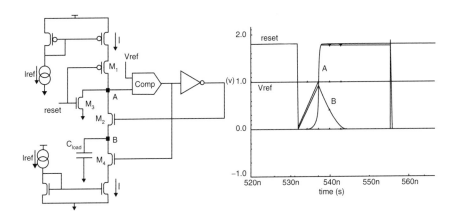

Fig. 7.6. The triangular pulse generator circuit

modulator. Here, node B of Figure 7.6 is the integrating node. However, in this case a discharging current source is also connected to node B. The switching transistors M2 and M4 connect the integrating node to the charging and discharging current sources respectively. The comparator inverts the state of these two transistors, hereby inverting the current flowing through capacitor C_{load}. Circuit simulation waveforms are given in Figure 7.6.

The Ring Oscillator and its Activation Circuit

The ring oscillator activation circuit uses the same principle as the PPM. By making its charging current twice the one required in the triangular pulse generator, the ring oscillator operates during the whole duration of the pulse. The carrier generator is a three-stage differential ring oscillator as shown in

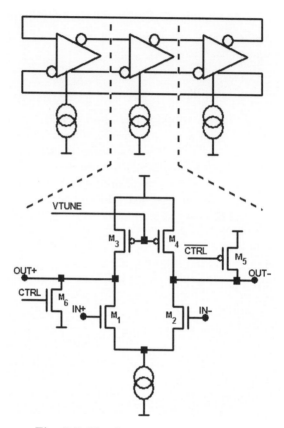

Fig. 7.7. The three-stages ring oscillator

Figure 7.7. Ring oscillators typically have very low startup times due to their low quality factors. Transistor $M5$ and $M6$ are used to force initial conditions on the internal nodes of the oscillator. By forcing initial conditions on each stages of the ring oscillator, the startup time is further decreased. In this way, the oscillator reaches its steady state in less than two cycles.

The Multiplier

The multiplier consists of a switching differential pair whose current source is modulated by the triangular waveform. This operation is realized with the circuit of Figure 7.8. The triangular voltage is applied to the gate of transistor M1. The resulting current is calculated from the large signal characteristic of an NMOS differential pair as follows:

$$I_{M1} = \frac{I_{ref}}{2} \pm \frac{I_{ref}}{2} \sqrt{\frac{\beta}{I_{ref}}(V_{in} - V_{b1})^2 - \frac{\beta^2}{I_{ref}^2}(V_{in} - V_{b1})^4} \qquad (7.4)$$

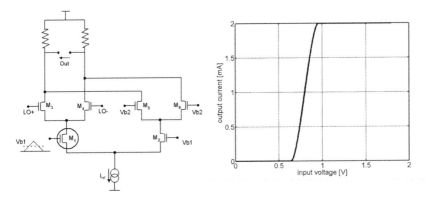

Fig. 7.8. Multiplier circuit consisting of a switching differential pair whose tail current is modulated by the triangular waveform

I_{ref} is the tail current of the differential pair and $\beta = \mu C_{ox} \frac{W}{L}$. This equation is plotted in Figure 7.8 for $I_{ref} = 2mA$, $V_{b1} = 0.8V$ and $\beta = 0.16\frac{A}{V^2}$. The output current pulse resulting from a triangular voltage of 0.9V peak and 6.2ns duration is plotted inside Figure 7.9. As long as transistor M1 does not cut M2 off, the voltage V_{b1} controls the amplitude of the current pulse. This current pulse, which is almost triangular, forms the envelope of the final output waveform. Indeed, the drain current of M1 flows through the switching differential pair M3-M4 producing an upconverted triangular pulse at the output. The current flowing in the path of transistor M2 ensures the stabilization of the DC voltage at the output. Since the waveform of the drain current of M1 defines the bandwidth of the pulse, it is important to test the variation of the pulse bandwidth with process variations. A process variation in the V_T is overcome by the use of a differential structure as can be seen in Equation (7.4). A mismatch in the V_T can be seen as an offset at the input of the differential pair. This offset can be directly removed by the control voltage V_{b1}. The effect of this offset is similar to a variation in the control voltage V_{b1} and will directly affect the amplitude of the pulse. However, equation (7.4)still shows a dependence of the current on β. The Fast Fourier Transform (FFT) of the current pulse is then taken to simulate the effect of β variations on the bandwidth. The result of the FFT is plotted on Figure 7.9. The bandwidth varies by 4% (20MHz for a 500MHz bandwidth) for a 20% variation in β.

Pulse Generator Measurement

The pulser circuit has been implemented in a logic 0.18μm CMOS technology [55]. The top picture of Figure 7.10 shows the measured spectrum of pulses with a bandwidth of 528MHz together with one corresponding time waveform in the top left corner. Three traces are depicted showing three different center frequency settings. The bottom picture of Figure 7.10 shows

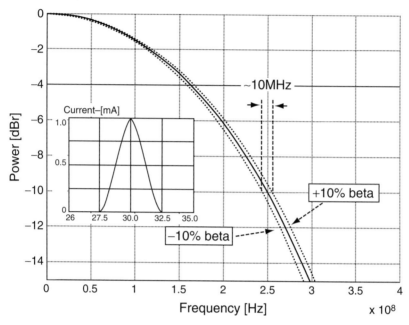

Fig. 7.9. Spectrum of the current pulse for ±10% variation of β

the 2GHz spectrum together with its time domain waveform. On top-left of of both graphs in Figure 7.10, the time-domain waveform are depicted for each of the bandwidth setting. A maximum output voltage of 200mV peak-to-peak is measured. These signals have been measured with a 20GS/s Tektronix TDS7404 oscilloscope. The burst power consumption is given in Figure 7.11 as a function of the pulse rate for pulse bandwidths of 500MHz, 1GHz and 2GHz. Below 10MHz repetition rate, the total power consumption is dominated by the baseline power consumption, which is slightly higher for the 2GHz pulses since the bias current has to be increased to compensate for the parasitic capacitance. This bias current is expected to decrease with technology scaling. Above 10MHz pulse repetition rate, the pulse creation dominates showing a linear increase of the power consumption with the pulse rate. The system reaches 50pJ per pulse energy consumption at a maximum of 40MHz repetition rate for a pulse bandwidth of 1GHz. For example, if ten pulses are used to code one bit, our pulser provides an average data rate of 10kbps with 5μW average power consumption.

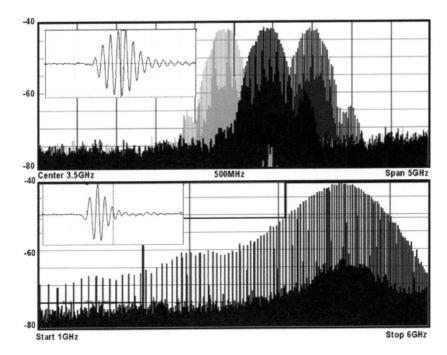

Fig. 7.10. Measured 528MHz (top) (with 3.432GHz, 3.960GHz and 4.488GHz center frequencies) and 2GHz (bottom) spectra and time waveform

7.3.2 UWB Receiver Design

UWB Receiver Architecture

Processing wideband analog signals in the digital domain requires an extremely fast sampling ADC with a wide input bandwidth. Such solutions [56] have all flexibility in terms of digital signal processing but are often damped by the required power consumption. Therefore, analog pre-processing is an interesting alternative to reduce the overall sampling rate and power consumption. Of course, such solutions are usually sub-optimal and prevent the use of very complex modulation schemes, but they can be very well suited for low data rate applications where power consumption is the main target. In this section, we propose an analog-based correlation receiver architecture, well suited for low data-rate impulse-based UWB applications (Figure 7.12). Since the correlation is done in the analog domain, the accuracy in the ADC sampling instants is shifted to a precise timing for the template generation. Usually, in order to optimally receive an UWB signal, the incoming pulse must be correlated with a template signal that precisely coincides with the incoming waveform. However, as shown on Figure 7.12, the carrier-based UWB signal is first down-converted in quadrature baseband, the matched template must

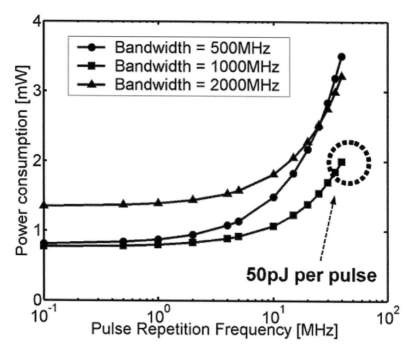

Fig. 7.11. Measured active burst power consumption as a function of the pulse repetition frequency

then correlate with the down-converted envelope of the pulse. Thereby, any timing inaccuracy is translated into a phase shift in the complex plane and almost no information is lost. The phase shift can be processed in the digital baseband to track the timing inaccuracy. The matched filtering is achieved through an analog integration operation over a precise time window. The time window is defined by the duration between the start of the integration process and the sampling instant. This operation corresponds to the correlation of the incoming triangular pulse with a rectangular window. The choice of such a simple receive template is driven by the simplicity of the implementation. The loss of signal energy can be calculated from the cross-correlation function between a rectangular function and a triangular function. Figure 7.13 shows the theoretical cross-correlation as a function of the rectangular window length normalized to the triangle length. The optimum corresponds to window of length 2/3 of that of the triangle, with a correlation of 0.9428 (0.5 dB loss from unit correlation of a perfect matched filter). The quadrature receiver can be used for both coherent and non-coherent modulation schemes. In the case of BPSK, the cross-correlation quadrature component have to be coherently combined from both branches. In the case of of PPM modulation, both branches are used to extract all the energy from the signal. On top of that,

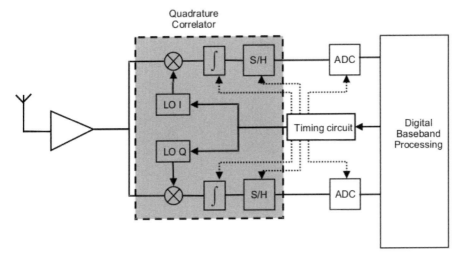

Fig. 7.12. Carrier-based UWB receiver

in the case of PPM modulation, a correlation must be done at each possible position of the pulse (2 for the case of binary PPM). Let's call I_1, Q_1 the quadrature samples at position 1 of PPM and, I_2, Q_2 the samples taken at position 2. For BPSK, only one position is needed (number 1). The decision variable $d_{pulse,i}$ for the pulse number i is then given by:

$$d_{pulse,i} = \begin{cases} I_1 \cos\phi + Q_1 \sin\phi & \text{for BPSK} \\ (I_1^2 + Q_1^2) - (I_2^2 + Q_2^2) & \text{for PPM} \end{cases} \quad (7.5)$$

where ϕ is the carrier phase of the received BPSK pulse with respect to the I branch. and that must be estimated through synchronization. There is a theoretical disadvantage of using non-coherent energy detection in the receiver, such as in the case of PPM modulation. However, the use of quadrature down-conversion in the receiver offers a benefit to PPM modulation that should not be neglected. Since the total complex envelope is retrieved from the signal, the synchronization accuracy is determined by the slowly varying envelope of the signal, which is used in case of energy detection. On the other hand, BPSK reception requires a perfect knowledge of the carrier phase, and any slight difference might ruin the detection process. For example, the reception of a 500MHz bandwidth pulse around 4GHz requires a timing accuracy that is 40 times more stringent for BPSK than for PPM. This particular example is illustrated on Figure 7.14, in order to keep at least 2/3 of the maximum correlation peak. This sensitivity to timing offsets is at the core of synchronization problems for UWB, already difficult for PPM, and much more for BPSK.

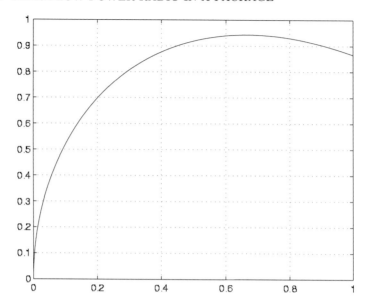

Fig. 7.13. Cross-correlation between a triangular shape and a rectangular shape. X-axis gives the rectangle length normalized to the triangle length, and Y-axis the correlation between both shapes

UWB Receiver Design and Measurements

The receiver architecture of Figure 7.12 has been implemented on a $0.18\mu m$ CMOS technology [54]. The front-end comprises an LNA and two quadrature down-conversion mixers (I and Q path) whose circuit is shown on Figure 7.15. In order to reuse current bias, the LNA and the mixers are stacked in a single structure. The circuit provides a differential 50Ohm input matching from 3GHz to 5GHz through two consecutive LC-stages. A cascode transistor isolates the LNA from the I and Q switching differential pairs. The LNA-mixer combination provides less than -10dB input reflexion coefficient in-band and 24dB gain with a 7.5dB cascaded noise figure over the whole band. An on-chip free-running tunable three-stages ring oscillator identical to the one used in the pulse generator in Section 7.3.1 provides the local oscillator (LO) for downconversion. The three 120 degrees output of the ring stagess are recombined into quadrature LO tones. The I and Q baseband outputs of the mixers are than filtered for channel selection through a 3rd order active filter [57] with variable gain (nominal value 0dB-to-15dB) and variable bandwidth (nominal value 250MHz-to-1GHz). It exploits the low linearity requirement in using a stacked MOS-C structure which allows to synthesize in a single path a 3rd-order transfer function. This features a low power consumption which ranges from $210\mu W$ (for 0dB gain and 250MHz bandwidth) to 3.2mW (for 15dB

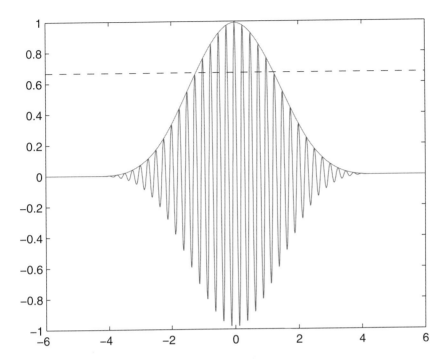

Fig. 7.14. Evolution of the normalized cross-correlation between transmit and receive pulses with time offset [ns]. The envelope corresponds to non-coherent reception, and the oscillating curve to coherent reception. A correlation equal to 2/3 is achieved for an offset of 1.2 ns and 33 ps respectively

gain and 1GHz bandwidth). The integrator circuit uses a positive feedback in the load to feature a dc-gain in excess of 60dB with a unity gain bandwidth of 1GHz, while consuming 120μW. No stability issue is present since it is operating in open loop configuration. The signal output is sampled by a differential 4-bit ADC consisting of 15 comparator stages, a set-reset latch array and a read-only memory (ROM) based Gray level encoder. Significantly downsizing the transistors and using calibration to cope with the increased mismatch errors have reduced its power consumption [58]. A SNDR higher than 22dB is measured up to 700MHz. This 3.5-ENOB ADC consumes a measured 1pJ/conversion step when running at 80 MS/s. A timing circuit (TC) generates the multi-phased signals that enable/disable the operation of the analog blocks. This TC sets the required time window to enable the operation of the analog block under consideration within the pulse frame. The input to the TC is the system clock, synchronized to the pulse repetition frequency.

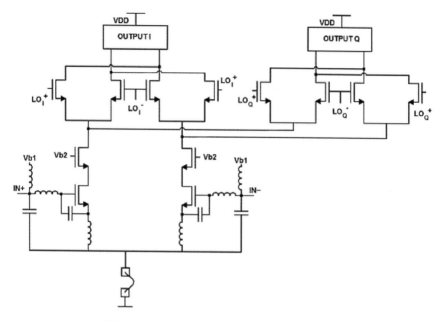

Fig. 7.15. Stacked LNA-Mixer combination

The analog outputs from the integrator are shown on Figure 7.16. The integrated correlation energy is shown on the top of the figure for a pulse stream modulated in PPM. The bottom of the figure shows a zoom on two consecutive pulses. The total current consumption of the chip including the digital baseband is 16mA measured on a 1.8V supply at 20MHz clock rate.

7.4 Heterogeneous Integration

7.4.1 Introduction

A sensor module (Figure 7.17) consists of several basic building blocks where each block has a clearly defined function within the module's operation. These blocks are:

- sensors and other environmental input
- actuators and other environmental output
- central processing
- wireless communication
- power management

Since applications drive specifications, the integration of a sensor module has to be designed with modularity and multi-application coverage in mind. The

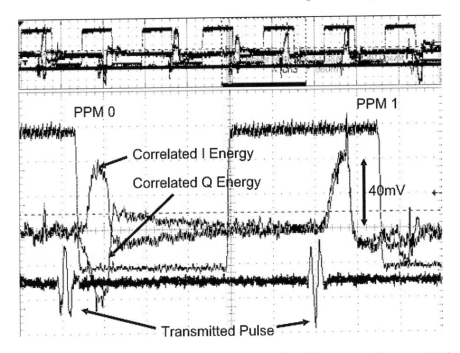

Fig. 7.16. Correlated output before ADC sampling. The consecutive correlated pulses are shown at two different PPM instants with respect to the reference clock

modularity is achieved by a layered design approach. Figure 7.18 gives an overview of the layered design approach.

7.4.2 Layered Design Approach

Wireless Communication Layer

The top layer is responsible for the wireless communication (Figure 7.19). It consists of a radio transceiver, a PCB integrated antenna, a 16MHz crystal and discrete passive components. For demonstration purpose, a Nordic NRF2401 transceiver has been used, but could further be replaced by any other radio such as the UWB radio described in the previous sections. The Nordic NRF2401 transceiver was selected here because of its low power characteristics. The NRF2401 features an energy per transmitted bit of 18nJ/bit. Operating in the 2.4GHz band, a 1MBit/s throughput at physical layer can be achieved with a maximum of 0 dBm output power. Due to MAC processing, packetising overhead and inter-chip communication between transceiver and microcontroller, the effective measured datarate between two nodes is set to a maximum of 400Kbit/s. The custom folded dipole antenna, integrated on

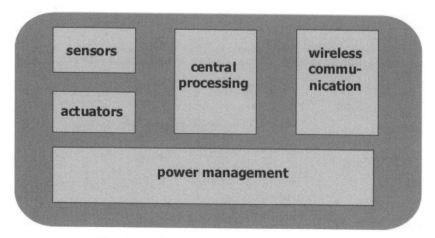

Fig. 7.17. General sensor node architecture

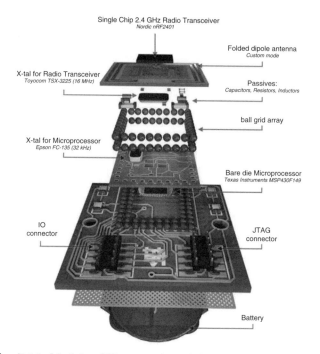

Fig. 7.18. Modular SiP approach with layered approach

the PCB, is a custom coplanar design and a trade-off between antenna size (miniaturization) and antenna efficiency.

Fig. 7.19. Uncut 3x3 PCB matrix, communication layer

Computation Layer

The second layer implements data processing and central control function-ality(Figure 7.20). It consists of a TI MSP430 microcontroller, a 32KHz crystal and passives. The TI MSP430 microcontroller was selected because of its market leading low power consumption. With a low active power (0.6nJ/instruction), a low standby power (2μW) and a fast wakeup from standby to active mode (6μs), the MSP430 is the ideal choice when low power is targeted. To meet the miniaturization criteria, a bare-die form factor was integrated by wire-bonding it directly to the printed circuit board and encap-sulating it by a glob-top coating. Interfacing to the outer world is achieved via the on-board peripheral blocks on the MSP430 microcontroller. An on-chip 12-bit analog to digital converter takes care of possible external analog sensors such as temperature, light, pressure or acceleration. The analog sensor output must reside in the 0 to 3V signal range. For digital inter-chip, sensor or actuator communication, standard peripheral interfaces such as parallel or serial port interface (SPI) can be used. In order to physically connect to the MSP430, these IO lines have been available on the battery layer through a small pitch connector. To program the on-board 60Kbyte FLASH memory, the JTAG interface is also made accessible through a connector on the battery layer. A cable from the PC parallel port to the JTAG connector provides a complete development chain for programming and debugging.

Fig. 7.20. Uncut 3x3 PCB matrix, computing layer with bare die MSP430

Battery and Power Management

The bottom layer is responsible for outside world interfacing and power management. It contains small pitch connectors, passives and battery. The IO connector is the main interface towards sensors and actuators; 8 input/output pins from the MSP430 are available. Every pin can be configured as a digital input/output or an analog input. The 10 pins IO connector also provides a ground signal and a 3V power signal; analog or digital sensors and actuators can be directly connected to the SiP module. The JTAG connector allows in-circuit programming and debugging of the MSP430 microcontroller. A wireless bootloader has also been developed in order to quickly update the SiP firmware without wire hassle. A standard 3V Lithium coin cell powers the SiP module. Lithium cells provide a high energy density (400Wh/L) which translates into a large capacity for a certain volume (200mAh for a CR2032 Lithium cell). Lithium cells also have a large peak current capability (rated 1C, 200mA peak). Since the peak current of the SiP can run up to 30mA (microcontroller active and RF active, no sensors), the Lithium cells provides backup to deliver the necessary current. Future developments will focus on an energy scavenging SiP without batteries. The sides of the SiP will be coated with solar cells with an efficiency of over 35% (Figure 7.21). The total area covered by the cells will be minimum 5cm^3. Since 100μW/cm^2 in average weather conditions can be achieved, a total average power of 500μW will be available through solar scavenging. Applications with a duty cycle of maximum 0.5% will be able to run in average conditions. For example, send temperature information every 10 seconds and go back to deep sleep. In order for the scavenging

Fig. 7.21. Future SiP coated with solar cells

module to "survive" at night and dark places, an energy storage unit can be provided, such as a gold cap or a rechargeable battery. Rechargeable batteries are preferred because of their larger capacity per volume figure. Li-Ion cells have a typical capacity of $0.1Wh/cm^3$. In order to survive a 12 hour period of darkness, an extra storage Li-Ion volume of $60\,mm^3$ is necessary for a $500\mu W$ application.

Interconnection

In order to achieve the 3D SiP form factor, a novel technique for layer interconnection was applied. The different functional layers (communication, computation, application) were designed on different boards and afterwards stacked on each other through a dual row of fine pitch solder balls. The solder balls have a 600 um diameter and placed with a 1.27 mm pitch. Each layer has a $14\,x\,14\,mm^2$ surface. Combining the communication and computation layer yields a functional module of $14\,x\,14\,x\,7\,mm^3$. The solder ball interconnect has multiple advantages:

- Each layer or module can be designed and tested separately
- Functional layers can be added or replaced, depending on the application
- Each layer can have different design and technology specifications, only the solder ball size/pitch/outline must match

The bottom footprint of each layer is a single row of solder balls, matching a standard BGA footprint. This allows each layer or a complete SiP to be placed in on a standard PCB with through standard BGA placement procedures.

7.4.3 Elementary Application

Hardware Description

Since the MSP430 has an on-board temperature sensor (1 degree accuracy), the most elementary demonstrator of the SiP is a wireless temperature sensor. In order to receive, plot and store temperature data from the sensor node, a basestation receiver with PC connection was required. Therefore, a basestation with USB stick form factor was developed (Figure 7.22). The stick basically has the same architecture and components of the sensor node (MSP430/nRF2401) but with a USB extension in order to communicate with a PC via a standard protocol. The USB chipset (FTDI component) operates according to the USB standard at physical level, but translates this to a RS232 device at application level, removing all USB burden for the application engineer. At PC side, a virtual COM port is created. The temperature

Fig. 7.22. USB stick basestation and SiP module

sensor demonstration is a duty cycled application. Every second, both the microcontroller and RF transceiver wake up from deep sleep. Then, a new

ADC sample from the on-board temperature sensor is taken. Together with some addressing bytes and frame bytes, the data is packetised and sent to the RF transceiver. The RF transceiver adds 1 byte CRC and transmits the total packet (80 bits in total) at 1Mbit/s. This yields an on-air time of 80 us. After transmission completion, both the microcontroller and RF transceiver go back to a deep sleep modes. The total active time is about 1 ms every second, yielding a 0.1% duty cycle. The average power consumption depends on duty cycle, sleep power consumption and active power consumption:

- active power consumption: 100 mW
- sleep power consumption: 5 μW
- duty cycle: 0.1%
- total average power consumption: 105 μW

The battery lifetime is determined by the total average power consumption. A CR2032 Li battery, used in the temperature sensor application, has a capacity of 600mWh. A 100μW application will thus last 6000 hours or 1.4 year.

MAC and Networking

The elementary temperature sensor application supports a simplex communication scheme: the sensor node is transmit only, the basestation is receive only. The sensor node sends ad-lib, without sensing for an occupied channel. Since temperature data is varying in the order of minutes, redundancy has been built in by sending the data every second, which allows a packet loss of more than 50%. Packet loss in a single node network can only be due to external interference sources, such as WiFi and Bluetooth, operating in the same 2.4GHz band. Adding multiple sensor nodes is possible, even with the same fire-and-forget MAC access scheme, better known as ALOHA. However, there is limit in the number of nodes in the ALOHA network to maintain a PER limit. When multiple nodes operate in a network and a minimum throughput per node has to be guaranteed, a beaconed TDMA access scheme is a reliable solution. The basestation periodically sends out beacons to all nodes. The nodes synchronize to the beacons and get a slot allocation for communication. One or more slots per node can be allocated; the larger the network, the smaller the allocation slot per node. If nodes should be able to dynamically join or leave the network, a join slot is preserved. A body area network has been built based on the TDMA access scheme (Figure 7.23). The network consisted of 2 sensor nodes: EEG/ECG (brain and heart activity) and EMG (muscle activity). The basestation sends a beacon every 10ms, followed by a join slot, allowing both nodes to join or leave ad-lib. A total of 5 communication slots can be allocated, with an application data size of 200 bits per slot. The total uplink application data rate is 100kbit/s. The EEG/ECG node takes 4 slots (80Kbit/s), the EMG node the remaining slot (20Kbit/s):

- EEG/ECG: 24 channels, 12-bit ADC resolution, 256 Hz sample rate = 74 Kbit/s
- EMG: 1 channel, 12-bit ADC resolution, 1 KHz sample rate = 12 Kbit/s

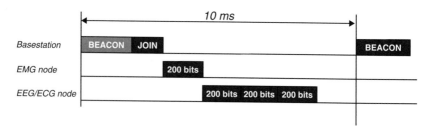

Fig. 7.23. TDMA access scheme for body area network

The beaconed TDMA access scheme requires both the nodes and basestation to apply transmit and receive functionality. A synchronized node wakes up from deep sleep a few milliseconds before the beacon is expected. After receiving the beacon, the node waits or goes back into sleep until its time slot has come. Then, the node sends a data packet. Since both nodes perform signal processing and on-board computation, the duty cycle for each node is over 80%, yielding a 80mW per node power consumption. Using the same lithium battery as with the temperature sensor, the battery only lasts 7 hours. No retransmission schemes are used in the BAN application; lost packets are interpolated if inter packet loss is minimal.

8

LOW ENERGY WIRELESS COMMUNICATION

By Ben Cook and Kris Pister. University of California, Berkeley.

8.1 Introduction

In this chapter, we will explore the energetic requirements of RF wireless communication from both a theoretical and practical standpoint. We focus on energy per transferred bit rather than continuous power consumption because it is more closely tied to the battery life of a wireless device. We begin with a look at the fundamental lower limit on energy per received bit imposed by the celebrated channel capacity theorem set forth by Claude Shannon. Based on this lower bound, we derive an energy efficiency metric for evaluating practical RF systems. By examining power-performance tradeoffs in RF system design, we begin to understand why and by how much will practical systems exceed this fundamental energy bound. From the discussion of system tradeoffs emerge a handful of low energy design techniques allowing systems to move closer to the fundamental energy bound. Finally, we present theory and measurements of a low energy 2.4GHz transceiver implemented in a 130nm RF CMOS process and discuss its energy saving architecture.

8.2 Fundamental Energy Requirements of Wireless Communication

Consider the task of properly detecting a signal with information rate R (in bits per second), and with continuous power P_0. The energy per bit in the signal is simply:

$$E_b = \frac{P_0}{R} \tag{8.1}$$

In this section, we use Shannons channel capacity theorem to determine the minimum value of E_b that will allow successful detection of the signal and relate this to other important system parameters. Shannons theorem establishes an upper bound on R for communication over a noisy channel. This bound is called the channel capacity C - in bits per second.

$$C = B \cdot log_2(1 + SNR) \tag{8.2}$$

B is the signal bandwidth and SNR is the ratio of signal power to noise power. If we assume the signal is corrupted by additive white Gaussian noise (AWGN), then 8.2 may be rewritten:

$$C = B \cdot log_2\left(1 + \frac{P_0}{N_0 B}\right) = B log_2\left(1 + \frac{E_b}{N_0}\frac{R}{B}\right) \tag{8.3}$$

N_0 is the noise power spectral density in Watts/Hz. P_0 is the signal power at the input of the receiver. If the channel is thermal noise limited, then N_0 is equal to the product kT, where T is temperature and k is Boltzmanns constant.

The ratio $\frac{E_b}{N_0}$ is referred to as the SNR-per-bit and the ratio $\frac{R}{B}$ is a measure of spectral efficiency in bps/Hz. Both quantities are important metrics for comparing digital modulation schemes. It is important to distinguish between SNR and $\frac{E_b}{N_0}$. SNR is a ratio of powers, while $\frac{E_b}{N_0}$ is a ratio of energies. For the purposes of evaluating a given schemes energy per bit performance, $\frac{E_b}{N_0}$ is more meaningful than SNR. For instance, if scheme A requires ten times greater $\frac{E_b}{N_0}$ for demodulation than scheme B, then scheme A will require ten times more energy to deliver a given data payload than B.

From 8.3, the capacity of a Gaussian channel increases logarithmically with signal power P_0. A cursory glance at 8.2 would suggest that C increases linearly with B, but the capacity-bandwidth relationship is actually more subtle due to the dependence of SNR on B. It turns out that C does increase monotonically with B, but only approaches an asymptotic value. Thus, for a given signal power P_0 and noise power density N_0, the channel capacity reaches its maximum value as B approaches infinity.

$$C_{max} = lim_{B \to \infty}\left(B log_2\left(1 + \frac{P_0}{N_0 B}\right)\right) = (log_2 e)\frac{P_0}{N_0} \approx 1.44\frac{P_0}{N_0} \tag{8.4}$$

Figure 8.1 and Figure 8.2 offer two different perspectives on Shannons theorem.

In Figure 8.1, the channel capacity is plotted versus signal bandwidth while P_0 and N_0 are held constant and in Figure 8.2, the maximum spectral efficiency (i.e. when R = C) is plotted against $\frac{E_b}{N_0}$ [59]. The minimum achievable $\frac{E_b}{N_0}$ follows from (1.4) by setting the information rate (R) equal to C_{max}.

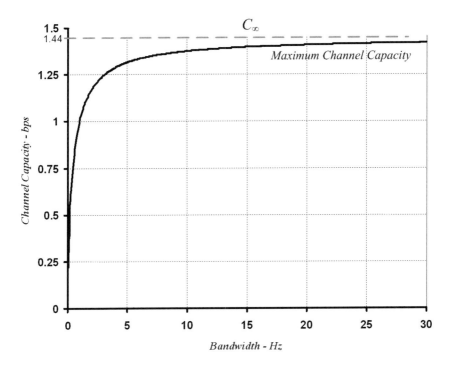

Fig. 8.1. Maximum achievable channel capacity as a function of bandwidth with constant $P_0/N_0 = 1$. $C_{max} = 1.44 \cdot P_0/N_0$

$$min\left(\frac{E_b}{N_0}\right) = \frac{P_0}{N_0 C_{max}} = ln(2) = -1.6 dB \qquad (8.5)$$

This powerful result tells us that error-free communication can be achieved so long as the noise power density is no more than 1.6 dB greater than the energy per bit in the signal. In a thermal noise limited channel (i.e. $N_0 = kT$) at 300K, the lower limit for Minimum Detectable Signal energy per bit (E_{b-MDS}) at the receiver input becomes:

$$min(E_{b-MDS}) = kT \cdot ln(2) \approx 3e - 21 \frac{Joules}{bit} \qquad (8.6)$$

Unfortunately, the theorem does not describe any modulation scheme that reaches the limit, and most popular schemes require far greater $\frac{E_b}{N_0}$ than - 1.6 dB. For a given modulation scheme (i.e. binary-PSK, OOK, etc.), the spectral efficiency R/B and minimum $\frac{E_b}{N_0}$ required for demodulation, call it $(\frac{E_b}{N_0})_{min}$, are fixed values, independent of transmission rate. The R/B and $(\frac{E_b}{N_0})_{min}$ values of the system may change, however, if coding is applied to the

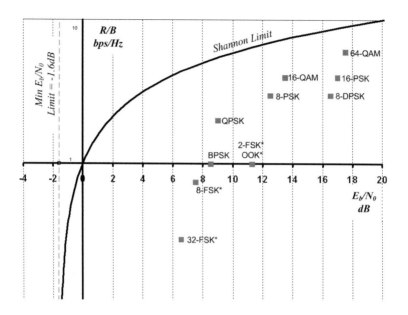

Fig. 8.2. Plot of maximum achievable spectral efficiency (R/B) versus required E_b/N_0 plus E_b/N_0 figures for several modulation schemes

modulation. Spread spectrum systems employ pseudo-noise (PN) codes that reduce R/B by increasing signal bandwidth (thus achieving processing gain), sometimes by several orders of magnitude, enabling reliable communication with SNR (power, not energy!) well below -1.6 dB. However, PN codes do not bring a system closer to the $\frac{E_b}{N_0}$ limit from (1.5), because the reduction in required SNR is compensated by the requirement of sending many chips per bit, so that overall energy per bit actually remains constant [60]. The purpose of PN codes is to spread the signal over a wider bandwidth, which is useful for: mitigation of multi-path fading, improved localization accuracy (i.e. GPS), multiple user access, interference avoidance, and more [59] [61]. Error correcting codes, on the other hand, can offer substantial reduction of energy per bit at the expense of system latency and computational power overhead.

8.2.1 Theoretical System Energy Limits

To this point, we have only considered the energy per bit at the input of a receiver. The goal is to find a lower bound on energy consumed by the system (including receiver and transmitter) per bit (E_{b-Sys}):

$$min(E_{b-sys}) = \frac{P_{TX} + P_{RX}}{R} \qquad (8.7)$$

P_{TX} and P_{RX} are the power consumed by the transmitter and receiver, respectively. In the best possible case, with a 100% efficient transmitter and zero power receiver, all the energy consumed by the system would go into the transmitted signal. Therefore, the fundamental lower bounds on E_{b-Sys} and transmitted energy per bit E_{b-TX} are the same.

$$min(E_{b-sys}) = min(E_{b-TX}) \qquad (8.8)$$

To find the lower bound on E_{b-Sys}, we now consider the minimum transmitted energy per bit E_{b-TX}. E_{b-TX} must exceed E_{b-MDS} to compensate for attenuation of the signal as it propagates from transmitter to receiver, or path loss. Path loss for a given link is a function of the link distance, the frequency of the signal, the environment through which the signal is propagating, and other variables. Accurate modeling of path loss is beyond the scope of this chapter, but a review of some popular models is offered in [62]. The ratio by which E_{b-TX} exceeds E_{b-MDS} is known as link margin (M) and is usually expressed in dB. For a reliable link, the system must have more link margin than path loss. In a thermal noise limited channel, the fundamental lower bound on E_{b-TX}, and thus E_{b-Sys}, required to achieve a link margin M is:

$$min(E_{b-sys}) = min(E_{b-TX}) = M \cdot KT \cdot ln(2) \qquad (8.9)$$

To achieve link margin M while only consuming $M \cdot kT \cdot ln2$ Joules per bit, a system must meet the following criteria:

- The receiver adds no noise
- The modulation scheme achieves the Shannon limit of -1.6dB for E_b/N_0
- The transmitter is 100% efficient
- The receiver consumes zero energy per bit

Clearly, such a system is impossible to design. In real systems, especially low power systems, transmitters are far from 100% efficient, the modulation scheme requires more E_b/N_0 than the limit, and the receivers are noisy and may consume a large portion of the total system energy. It is not uncommon for a system, especially a low-energy system, to consume 10,000 times more energy per bit than this lower limit. For instance, radios targeting sensor network applications have reported link margin of 88-120dB [8] [36] [9] [7] [63] [64] [65] [66] [67], resulting in a theoretical minimum energy per bit of 1.9-3000 pJ, but the actual energy consumed by these systems per bit ranges from about 4.4-1320 nJ. Since the lower bound on E_{b-Sys} scales with M, and M may vary over several orders of magnitude from system to system, a simple comparison of E_{b-Sys} is not really fair. To let us compare apples to apples, we define an energy efficiency figure of merit for communication systems with an ideal value of 1:

$$\eta = \frac{ideal\ energy/bit}{actual\ energy/bit} = \frac{M \cdot KT \cdot ln(2)}{E_{b-sys}} \tag{8.10}$$

The η values for the sensor network radios previously mentioned are shown in Figure 8.3.

Standard	Ref. #	Frequency Band	Modulation Type	Data Rate Mbps	P_{RX} mW	P_{TX} mW	Link Margin dB	Energy F.O.M. η dB
WLAN 802.11G	[13]	2.4 GHz	64-QAM (OFDM)	54	1320	2145	95	-38
		2.4 GHz	BPSK (OFDM)	6	1320	2145	115	-28
802.15.4	[8]	2.4 GHz	O-QPSK	0.25	26.5	28.3	105	-34
	[12]	2.4 GHz	O-QPSK	0.25	30.1	27.8	95	-44
Bluetooth 802.15.1	[15]	2.4 GHz	2-GFSK	1	70	49	85.5	-51
N/A	[9]	2.4 GHz	2-FSK	0.3	0.33	1	92	-30
	[5]	1.9 GHz	OOK	0.005	0.4	1.2	96.3	-44
	[6]	900 MHz	2-FSK	0.1	1.2	1.3	88	-41
	[4]	434-868 MHz	2-FSK	0.025	2.1	30	121	-25

Fig. 8.3. Energy efficiency data for various commercial and academic radios

We have mentioned several factors contributing to low energy efficiency in wireless systems. Now our goal is to capture the relative impact of said factors by incorporating them into an expression for η. We begin by redefining link margin.

$$M = \left(\frac{E_{b-TX}}{E_{b-MDS}}\right) = \frac{E_{b-TX}}{F \cdot KT \cdot (E_b/N_0)_{min}} \tag{8.11}$$

$(\frac{E_b}{N_0})_{min}$ is the minimum SNR-per-bit required for demodulation and F is called the receiver noise factor. F is a non-ideality factor (F = 1) for the noise performance of a receiver and is discussed in greater detail in Section III. In the ideal case, F = 1 and $\frac{E_b}{N_0} = ln(2)$. Equation 8.11 allows us to express η in a much more intuitive form.

$$M = \left(\frac{E_{b-TX}}{E_{b-sys}}\right) \cdot \frac{1}{F} \cdot \left(\frac{ln(2)}{(E_b/N_0)_{min}}\right) \tag{8.12}$$

Each of the three terms in 8.12 may assume values from 0 to 1 and has an ideal value of 1. The first term tells us what portion of the total energy consumed by the overall system gets radiated as RF signal energy in the transmitter. The second term describes how much the link margin is degraded due to noise added by the receiver. The third term quantifies the non-ideality of the systems modulation/demodulation strategy as compared to the minimum achievable $\frac{E_b}{N_0}$ from 8.5.

Wireless systems with very high output power tend to have higher η because transmitter overhead power and receiver power do not scale up with transmitted power; a larger proportion of the overall power budget will burned in the PA. This is evident in Figure 8.3 where the 2 highest η values come from the systems with highest output power. For this reason, it is most useful to compare η for systems with similar output power.

Equation 8.12 provides a good starting point for further exploration of low energy system design, but it is not a perfect metric and there are a few caveats attached with its use. First of all, we have not considered dynamic effects such as the "startup energy" spent as the transceiver tunes to the proper frequency. Nor have we included network synchronization or the overhead bits due to training sequences, packet addressing, encryption, etc. Rather than attempt to capture all the initialization effects that lead to radios being on with no useful data flowing, we have narrowed our scope by assuming the transmitter and receiver are already time synchronized and their typical data payload per transmission is large enough that startup energy is negligible. At this point, we shift our focus to design of low-energy wireless communication systems and discuss techniques that can improve η.

8.3 Low Energy Transceiver Design

In the discussion that follows, we examine the impact of modulation scheme on system energy consumption and transceiver architecture and then discuss general design techniques for boosting transmitter efficiency and building low noise, low power receivers.

8.3.1 Modulation Scheme

Modulation scheme directly impacts a communication systems bandwidth efficiency (R/B) and minimum achievable energy per bit ($\frac{E_b}{N_0}$). A reasonable question to ask is: which has the potential for lowest energy per bit, a complex modulation scheme that packs many bits of data into each signal transition, or a simple binary scheme? The answer is not obvious because there is a trade-off; more complex schemes achieve higher information rates but typically also require higher SNR to demodulate.

Figure 8.2 provides a comparison of several popular modulation schemes with respect to the Shannon limit, plotting R/B versus the $\frac{E_b}{N_0}$ required for reliable demodulation. If system link margin is held constant, then the best modulation strategy will largely be determined which resource is more precious, bandwidth or energy. Schemes with lower $\frac{E_b}{N_0}$ will deliver more data for a fixed amount of energy, while those with higher R/B will deliver highest transmission rate for a fixed amount of bandwidth.

As an example, the 802.11g standard employs 64-QAM (OFDM on 48 sub-carriers) to achieve 54Mbps in the crowded 2.4GHz ISM band while only occupying about 11MHz of bandwidth. In the case of 64-QAM, high bandwidth efficiency comes at the cost of poor energy efficiency as evidenced by its high $\frac{E_b}{N_0}$ requirement. On the other hand, 802.11g specifies a 6Mbps mode which uses BPSK (OFDM on 48 sub-carriers) also occupying 11MHz and having the same coding rate as the 54Mbps mode. Using BPSK, the data rate only decreases by a factor of 9 but the 802.11 spec requires a 60X receiver sensitivity improvement over the 54Mbps mode, owing to the lower $(\frac{E_b}{N_0})_{min}$ of BPSK versus 64-QAM. Figure 8.3 provides sensitivity, power consumption, link margin, and η data from an 802.11G chipset using these modulation methods [68].

Since we are most concerned with minimizing energy consumption, we would tend to favor a modulation scheme with as small an $(\frac{E_b}{N_0})_{min}$ requirement as possible. Furthermore, given the relatively low data throughput and short range of the systems of interest, some sacrifice of bandwidth efficiency is justifiable if it affords an energy benefit. In theory, the lowest energy uncoded modulation scheme would be M-ary FSK with M approaching infinity [59]. This strategy is not popular because $(\frac{E_b}{N_0})_{min}$ only decreases incrementally at large M, while the occupied bandwidth and system complexity grow steadily.

In practical systems targeting low energy, 2,4-PSK, 2-FSK, and OOK are the most common modulation methods - representing a compromise between energy efficiency and simplicity of implementation. Radios designed for sensor network applications have used either PSK [63] [67], binary FSK [8] [9] [7] [64] [65] [66], or OOK [8] [36]. The original 802.15.1 standard (Bluetooth) uses Gaussian 2-FSK and the 802.15.4 standard uses a form of QPSK (i.e. 4-PSK) that can be implemented as 2-FSK. Newer versions of Bluetooth adopt 8-DPSK as the modulation technique to extend data rate to 3Mbps, but the energy efficiency η of these systems will most likely drop somewhat $(\frac{E_b}{N_0})_{min}$ for 8-DPSK is substantially higher than the original GFSK format.

System Architecture Considerations

When choosing a modulation scheme for low-energy, $(\frac{E_b}{N_0})_{min}$ does not tell the complete story. Even if $(\frac{E_b}{N_0})_{min}$ is low, the overall system can still be inefficient if the power needed to generate, modulate, and demodulate the signal is comparable to or larger than the transmitted power. For applications requiring relatively small link margin (i.e. low transmit power), such as WPAN and sensor networks, it becomes particularly important to choose a modulation scheme that requires little power to implement so that the system may remain efficient even with low power output. An ideal modulation scheme would maximize link margin or capacity for a given signal power (i.e. smallest $(\frac{E_b}{N_0})_{min}$) without requiring complex, high-power circuits.

802.11g in its highest data rate represents a good example of "what not to do" if energy conservation is the goal because 64-QAM has a high $(\frac{E_b}{N_0})_{min}$ and its implementation is generally power hungry and quite complex. The receivers are high power because demodulation requires a fast, high-precision ADC, substantial digital signal processing, and linear amplification along the entire receive chain. The 802.11g transmitters tend to be power hungry because generating the 64-QAM signals requires a linear PA and a fast, low-noise PLL and VCO. Since the transistor devices constituting the amplifiers (and all blocks) in a transceiver are inherently nonlinear, achieving linear amplification in the receive chain and PA comes at the cost of increased power and/or complexity.

In contrast to QAM and PAM, FSK and PSK have a common trait that only one nonzero signal amplitude must be generated. This has important consequences for system efficiency. First of all, the PA can be a nonlinear amplifier - making much higher efficiency possible. Secondly, since information is only carried in the phase (or frequency) of the signal, the receive chain need not remain linear after channel selection, so demodulation can be accomplished with a 1-bit quantized waveform. Finally, with FSK (and some forms of PSK) it is possible to generate the necessary frequency shifts by directly modulating the frequency of the VCO, thereby eliminating the transmit mixer and saving power.

The potential power savings of direct VCO modulation depend strongly on the phase accuracy required of the transmitter. If moderate frequency or phase errors are tolerable, the VCO can simply be tuned directly to the channel with a digital FLL and modulated open-loop [9] - resulting in a simple, low power implementation. For phase-error intolerant specifications such as GSM, a variant of direct VCO modulation known as the 2-point method is often used. In the simplest version of the 2-point method, a continuous time (fractional-N) PLL with relatively low bandwidth attempts to hold the VCO frequency steady while an external input modulates the VCO frequency. A high precision DAC feeds forward a signal to cancel the "error" perceived by the PLL due to the modulation [69]. Though the 2-point method eliminates the need for a transmit mixer, the power consumed by the DAC and PLL curtail the potential power savings somewhat. This method has been verified for 802.15.4 [63], Bluetooth [70], GSM [71], and other standards.

Error Correcting Codes (ECC)

With respect to modulation scheme, a tradeoff between spectral efficiency and energy efficiency has emerged from both theoretical and practical perspectives. First of all, Shannons capacity theorem shows that the minimum achievable energy per bit for any communication system is logarithmically related to spectral efficiency and several popular (uncoded) modulation schemes, though not approaching the Shannon limit, do exhibit a strong positive relationship between R/B and $\frac{E_b}{N_0}$. Further, from a practical perspective, the schemes with highest R/B, such as m-PAM or m-QAM with large m, require complex and high power hardware to implement. The confluence of these factors suggest that simpler schemes, such as 2-FSK, OOK, and 2,4-PSK, will offer the best tradeoff when minimizing energy is the goal.

Even with an optimal demodulator, 2,4-PSK, 2-FSK, and OOK still require at least 10 times higher $(\frac{E_b}{N_0})_{min}$ than the Shannon limit to achieve a low probability of error (i.e. BER = 10-5). The capacity equation (1.3) tells us that, to approach the Shannon limit and reclaim some of this wasted energy, the bandwidth efficiency R/B will have to be reduced. Error correcting codes (ECC), such as Hamming, Reed-Solomon, and Turbo Codes, can reduce $(\frac{E_b}{N_0})_{min}$ significantly, but also incur substantial computational power overhead that could increase E_{b-Sys} enough to outweigh the $(\frac{E_b}{N_0})_{min}$ reduction, particularly in low power systems. In [60], the $(\frac{E_b}{N_0})_{min}$ reduction (or coding gain) and digital computation energy of several ECCs were evaluated for a 0.18m CMOS process with 1.8V supply. See Figure 8.4.

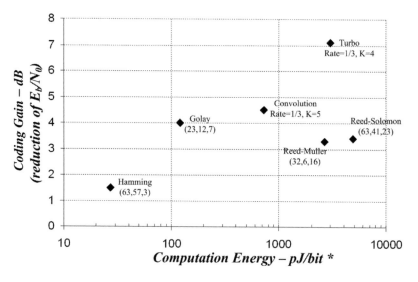

Fig. 8.4. Coding gain computational overhead in a $0.18\mu m$ CMOS process

Though ECCs have traditionally found use in higher power systems, these estimates would suggest that digital computation energy is now low enough that ECCs are an effective option. ECCs will only become more favorable as supply voltages and digital process features continue to scale.

8.3.2 Minimizing Overhead Power

The basic functions of the transmitter are: generate a stable RF signal, modulate the frequency, phase and/or amplitude of the RF signal according to information to be transmitted, and drive the modulated signal onto the antenna with a PA. In a sense, energy consumed by the modulation and signal generation circuitry constitutes overhead because it does not contribute directly to the systems link margin. This overhead power $(P_{OH,TX})$ is, to first-order, independent of transmitter output power. Efficient transmitter designs will spend proportionally small amounts of energy generating and modulating the RF signal, with the greatest share of energy consumed by the PA.

The receiver functions can be summarized as: low-noise, linear amplification, selection of communication channel, and demodulation. The low noise amplifier (LNA) boosts the incoming signal amplitude to overcome the noise of subsequent stages while adding as little noise and distortion as possible. The excess noise contributed by an LNA is inversely related to its power consumption; increasing power in the LNA directly increases link margin. Channel selection and demodulation are typically accomplished with a VCO, mixers, low frequency filters, and other circuits. A certain amount of power $(P_{OH,RX})$ must be spent in these blocks for the receiver to function, but increasing their power beyond that point does not have as direct an impact on link margin as increasing LNA power. A first order model of the power performance tradeoffs in a generic transceiver is illustrated in Figure 8.5 [2].

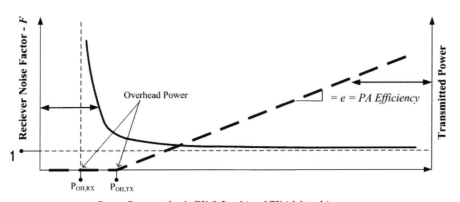

Power Consumption in RX (left-axis) and TX (right axis).

Fig. 8.5. Graphical representation of first order model of power-performance tradeoffs in an RF transceiver [2]

The overhead power $(P_{OH,RX}$ and $P_{OH,TX})$ spent generating and demodulating the RF signal is strongly dependent on the hardware requirements of the modulation scheme employed. From a hardware standpoint, the modulation schemes with lowest overhead are OOK and 2-FSK *(with large frequency deviations)* because they require only a single non-zero signal amplitude and are tolerant of moderate phase/frequency errors. These relaxed specifications permit simpler, lower power modulation and demodulation circuits so that a larger proportion of the overall power can be burned in the PA and LNA. However, even the most barebones low-IF or direct conversion transceivers still require an RF VCO to operate. Thus, *in the limit of system simplicity, overhead power is VCO power.*

Overhead Power in the VCO

A VCO is an autonomous circuit with either feedback or negative resistance designed to cause periodic oscillation at one frequency; that frequency is set by an RC, RL, or resonant RLC network. See Figure 8.6.

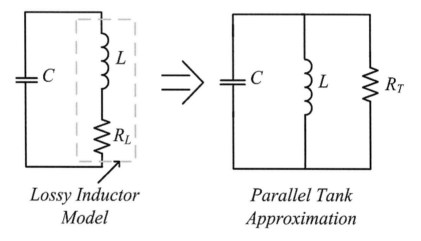

Lossy Inductor Model   *Parallel Tank Approximation*

Fig. 8.6. Model for LC resonator with loss dominated by inductor. Right: Parallel LC approximation with tank impedance RT at resonance

The vast majority of VCOs designed for communication systems use a parallel LC resonator (or LC tank) to select the frequency of oscillation because of its potential for superior noise performance. The power requirements and noise performance of an LC VCO are largely determined by the impedance at resonance (R_T) and quality factor (Q_{tank}) of this resonant LC tank. Integrated circuit processes are inherently better suited to making capacitors than inductors and, for frequencies below about 10GHz, the value of Qtank is usually limited by the losses in the inductor. The inductor quality factor (Q_L) is:

$$Q_L = \frac{\omega_0 \cdot L}{R_L} \approx Q_{tank} \qquad (8.13)$$

For the parallel LC tank in Figure 8.6 (left), the approximate magnitude of the tank impedance at resonance (R_T) is given by:

$$R_T = \omega_0 \cdot L \cdot Q_L \qquad (8.14)$$

Some popular LC VCO topologies are shown in Figure 8.7 [72]. A certain amount of current is needed for oscillation to begin, but the current required to meet output swing requirements is usually much greater. Typically, Vo must be at least a few hundred milliVolts. Vo can be expressed as a constant times the product of I_{SS} and R_T for both VCOs in Figure 8.7. Hence, R_T must be maximized to minimize current, making high value, high-Q inductors critical to reducing power in the VCO.

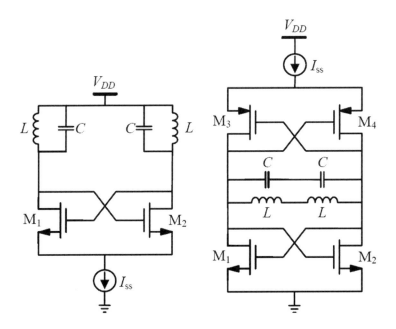

Fig. 8.7. Two popular negative resistance LC VCOs. The NMOS only VCO only delivers half the output swing per current of the complementary VCO (right), but its maximum achievable swing is twice as large

The choice of VCO topology is also an important consideration for minimizing power. For instance, Vo as a function of I_{SS} and R_T for the NMOS only circuit is [72]:

$$V_0 \approx \frac{2}{\pi} \cdot I_{ss} \cdot R_T \tag{8.15}$$

Whereas, V_o for the complementary (CMOS) VCO is:

$$V_0 \approx \frac{4}{\pi} \cdot I_{ss} \cdot R_T \tag{8.16}$$

The CMOS VCO will deliver twice the output swing for a given current, its maximum swing is just hal that of the NMOS only device, which swings about the supply rail. Thus, the CMOS VCO would be the preferred choice unless it can not achieve sufficient swing. The CMOS VCO provides greater swing per current because the commutating current ISS flows through a parallel impedance of $2R_T$, whereas the impedance seen by I_{SS} in the NMOS VCO is only R_T. The CMOS VCO can also be seen as a vertical stack of two VCOs sharing the same bias current. As well see below, stacking RF circuits to reuse bias current is a powerful tool for improving system efficiency.

Voltage Headroom and RF Circuit Stacking

Even with the most barebones transceiver architecture significant power may still be wasted if the available voltage headroom is not used optimally. Many mobile systems use a 3.3V lithium supply, but the voltage swing required by the PA, VCO, or LNA may be much lower. For instance, if a VCO is powered by a 3.3V supply but only needs to generate a $300mV_{0-pk}$ signal to drive mixers, buffers, or frequency dividers, there will be substantial waste because the VCO swing specification could be met with a much lower supply voltage.

Since supply voltage is typically not a flexible design variable, circuit techniques are needed to optimize use of headroom when supply voltage is high. One way to reduce wasted power is by stacking RF circuits [9]. Stacking is accomplished by placing two RF blocks in series with respect to bias currents flowing from the supply. Thus, the current used in one block is reused by another block. To avoid signal crosstalk between the two blocks, they are isolated from each other with a large decoupling capacitor that provides a low impedance node at high-frequencies. For integrated transceivers, stacking is only feasible for high-frequency circuits where effective isolation can be implemented with on-chip decoupling capacitors.

A few different stacked configurations are shown in Figure 8.8.

The effect of stacking two small-signal LNAs is to either double the transconductance gm (if the inputs and outputs are coupled in parallel), or to increase the voltage gain Av (if signals traverse the LNAs in series). Stacking two PAs doubles the output current, provided the halved voltage headroom is still sufficient. PA stacking techniques are discussed in more detail in Section 8.3.4. In [9], the VCO was stacked with the LNA in the receiver and with the PA in the transmitter. In this design, the current available to the PA and LNA was set by the VCOs current requirements.

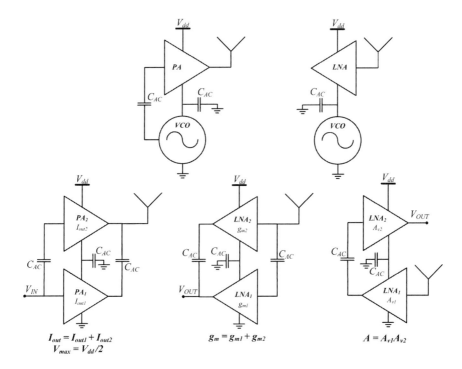

Fig. 8.8. Various stacked RF circuit configurations to share bias current and optimize voltage headroom usage

8.3.3 Receiver Noise Factor and Passive Voltage Gain

In this section, we will see how high impedance and passive voltage gain allow a receiver to achieve good noise performance with reduced power. Noise performance of RF receivers is most often reported using the noise factor (F) - defined as the ratio of the SNR at the receiver input to the SNR at the output. From a system perspective, F is the factor by which link margin is degraded by the receivers own internal noise generators. To maintain a given link margin, an increase in F must be compensated by an equivalent increase in transmitted power. In the absence of an input signal, F can be expressed as the ratio of the systems total output noise to the output noise due to the source resistance. Referring to stage S1 with voltage gain A_v in Figure 8.9 (left), the squared voltage noise at the output is the sum of the source noise times $|A_v|^2$ and the noise added by S1.

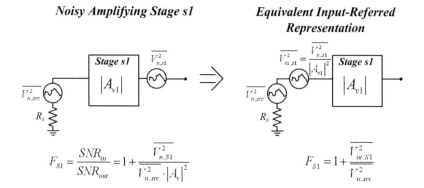

Fig. 8.9. Left: Noise factor calculation for voltage amplifying stage s1 with source noise due to R_s. Right: Input-referred representation of s1

Thus, F can be expressed:

$$F_{S1} = \frac{SNR_{in}}{SNR_{out}} = \frac{\overline{V_{n,S1}^2} + \overline{V_{n,src}^2} \cdot A_v^2}{\overline{V_{n,src}^2} \cdot A_v^2} \tag{8.17}$$

Without loss of generality, we have chosen to sum noise contributions using voltage gains and squared voltage noise rather than power gain and noise power. Summing noise voltage is more convenient when the impedance between stages within the receiver are not specified - which is typically the case in integrated transceivers. We add rms voltage noise because we assume the noise sources are uncorrelated. Alternatively, we can represent the noise added by S1 with an equivalent input voltage source that produces the same total output noise (Figure 8.9, right).

$$\overline{V_{ni,S1}^2} = \frac{\overline{V_{n,S1}^2}}{A_{v1}^2} \tag{8.18}$$

V_{ni}^2 is called the input referred noise voltage of S1. Referring noise to the input is useful for determining minimum detectable signal levels because it gives a direct measure of how large an input signal must be to overcome the noise contributed by the system and source noise. From 8.18, we can express the noise factor of S1 in terms of its input referred voltage noise.

$$F_{S1} = \frac{SNR_{in}}{SNR_{out}} = \frac{\overline{V_{n,S1}^2} + \overline{V_{n,src}^2}}{\overline{V_{n,src}^2}} = 1 + \frac{\overline{V_{n,S1}^2}}{\overline{V_{n,src}^2}} \tag{8.19}$$

A receiver is a cascade of stages, each having a different voltage gain and noise contribution Figure 8.10. Each stage amplifies the signal and noise at its input and adds its own noise.

Cascade of Noisy Amplifying Stages

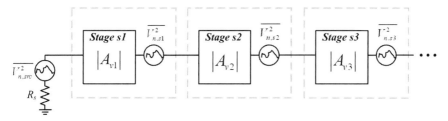

Input-Referred Representation and Cascaded Noise Factor

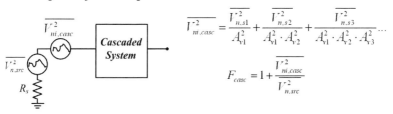

Fig. 8.10. Cascade of amplifying stages with uncorrelated noise sources modeling a receive chain. Bottom: Input-referred representation and noise factor

In general, the noise added by each stage is uncorrelated with the signal at its input. If A_{vk} and $V_{n,k}^2$ represent the gain and output noise of the k_{th} stage, respectively, then the noise factor the cascaded system can be expressed.

$$F_{casc} = 1 + \frac{1}{V_{n,src}^2}\left(\overline{V_{ni,1}^2} + \frac{\overline{V_{ni,2}^2}}{A_{v1}^2} + \frac{\overline{V_{ni,3}^2}}{A_{v1}^2 \cdot A_{v2}^2} + \ldots\right) = 1 + \frac{\overline{V_{ni,casc}^2}}{V_{n,src}^2} \quad (8.20)$$

8.20 shows that the impact of noise added by a given stage is reduced by the square of the total voltage gain preceding it. Typically, the first active stage in a receiver is a low-noise amplifier (LNA) achieving roughly 15-25dB of voltage gain. Thus, the following stages can have much greater input referred noise than the LNA and still only a minor effect on the cascaded system noise factor.

The noise contribution of an LNA depends on the current consumption, device technology, circuit topology and other factors. However, LNA noise is usually dominated by the input transconductor, consisting of one or more transistors biased for small signal amplification. The input referred voltage noise of a CMOS transconductor (or Bipolar device) can be related to current consumption directly:

$$\frac{\overline{V_{ni}^2}}{\Delta f} = 4kT\frac{\gamma}{g_m} = 2kT\frac{\gamma \cdot v_{dsat}}{I_d} \tag{8.21}$$

V_{dsat} is called the saturation voltage and the right side of 8.21 holds (roughly) for $V_{dsat} = 100$mV. Though 8.21 just represents the input noise of a single MOS transistor, its basic form is common to most LNA topologies. Furthermore, the input referred noise of mixers, low-frequency filters, and other stages following the LNA will generally be inversely related to current consumption in a similar manner. Hence, from 8.20 and 8.21, it is clear that voltage gain at the front of the receiver chain reduces the current required to meet a given noise specification.

Passive Voltage Gain with Resonant LC Networks

It is possible for a stage to achieve voltage gain without increasing the power in the signal. This is only possible when the impedance at the output is larger than at the input. For instance, if a given block is lossless and has an output impedance 100 times greater than the input, then the output voltage will be 10 times larger than the input (because power $= \frac{V^2}{R}$ will remain constant), but the signal current will decrease by a factor of 10.

Passive transformers, resonant LC circuits, or even resonant electromechanical devices, can achieve voltage gain while consuming zero power. Passive voltage gain is a powerful tool for reducing receiver power consumption and is particularly well suited to CMOS because CMOS transistors accept voltage (rather than power) as input and have a capacitive input impedance that can be incorporated into a resonant network without contributing much loss.

Figure 8.11 is a tapped-capacitor resonant transformer, an LC network capable of delivering passive voltage gain. In this circuit, R_S is the source resistance and R_L models resistive loss in the inductor. The quality factor of an inductor (Q_L) is a common metric that quantifies how close an inductor is to ideal. An ideal inductor would have infinite Q_L.

The source driving the RF port has magnitude $2V_i$ to account for the voltage dropped across R_S. If the input impedance of the network is matched to R_S, then $V_a = V_i$. Note that in reality there will be some capacitance in parallel with the inductor due to both finite inductor self resonance frequency (SRF) and the input capacitance of any devices connected to the output port. If the value of this parasitic capacitance is small relative to C_1, then we can safely neglect its impact.

From the output port, the matching network appears as a simple parallel LC tank with a lossy capacitor and inductor. The lossy capacitor consists

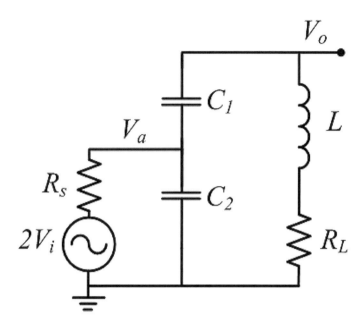

Fig. 8.11. Circuit model for tapped-capacitor resonator

of elements C_1, C_2, and source resistance R_S and its effective quality factor (Q_C) is set by R_S and the ratio of capacitors C_2 and C_1. The overall network Q at the parallel resonance is a parallel combination of Q_L and Q_C. Q_C may assume a wide range of values depending on the values of C_1 and C_2, permitting design flexibility. Q_C is defined:

$$Q_L = \omega_0 R_s \frac{C_2}{C_1}(C_2 + C_1) + \frac{1}{\omega R_s C_1} \tag{8.22}$$

The output impedance of the network at resonance is real and its magnitude is:

$$R_0|_{\omega=\omega_0} = \omega_0 L \frac{Q_C Q_L}{Q_C + Q_L} \tag{8.23}$$

Noise at the output is contributed by both R_S and R_L. If Q_C is very large, then the overall Q is limited by the inductor and most noise at the output will come from R_L, leading to high noise factor. On the other hand, the network has the lowest noise factor when Q_C is much smaller than Q_L because losses and output noise are dominated by R_S. To quantify the relationships between L, Q_L, and Q_C we first determine the voltage gain from noise voltage sources in series with both R_L and R_S to the output, denoted A_{VL} and A_{VS}, respectively:

$$|A_{VL}(\omega_0)| \approx \frac{2Q_C Q_L}{Q_C + Q_L} \tag{8.24}$$

$$|A_{VS}(\omega_0)| \approx \frac{2Q_C Q_L}{Q_C + Q_L} \sqrt{\frac{\omega_0 L}{R_s Q_C}} \tag{8.25}$$

Therefore, the noise factor (F) of the network at resonance becomes:

$$F|_{\omega=\omega_0} = 1 + \frac{R_L}{R_S} \left(\frac{|A_{VL}(\omega_0)|}{|A_{VS}(\omega_0)|} \right)^2 = 1 + \frac{Q_C}{Q_L} \tag{8.26}$$

The maximum gain is achieved when the source impedance is perfectly matched to R_L. This is an intuitive result because all power delivered to the network must be dissipated in R_L and the output voltage is largest when the current through R_L is maximum. Matching occurs when Q_L and Q_C are equal. Thus, from 8.26 the noise factor is 2 (NF = 3dB) when matched. The voltage gain of the network when matched is:

$$|A_{VS}(\omega_0)|_{max} \approx \frac{1\omega_0 L}{\sqrt{R_s R_L}} \tag{8.27}$$

The noise factor, gain and S_{11} of a tapped-capacitor network are plotted versus C_2 in Figure 8.12. In this example, the resonant frequency is 2.45GHz, inductance is 10nH and Q_L is 18. The inductance and Q_L are roughly based on values achieved with integrated inductors in a current 130nm RF CMOS process. Though the network increases the voltage amplitude of the signal, it actually decreases the signal power by a factor of F^{-1}. If used in a receiver front-end, as in [73], this network places a lower limit on the achievable system noise factor, but it consumes no power, remains perfectly linear and allows for substantial power reduction in subsequent stages due to its voltage gain.

8.3.4 Efficient PAs with Low Power Output

As output power creeps below 1mW or so, designing an efficient transmitter becomes increasingly difficult. First of all, there are numerous system blocks whose power consumption does not necessarily scale down with transmitted power - resulting in a proportionally large power overhead. Secondly, with typical supply voltages of 1-3V and an antenna impedance of roughly 50 Ω, standard PA topologies will be inherently inefficient when putting out such little power. We have seen how power overhead can be minimized by choosing the right modulation scheme, VCO topology, and, if necessary, RF circuit stacking. We will now address multiple techniques for increasing the efficiency of low power PAs. When transmitting a constant envelope signal, a nonlinear

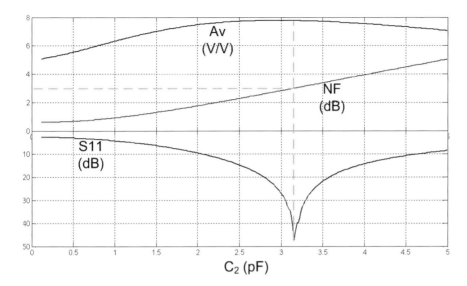

Fig. 8.12. Noise factor, voltage gain, and S_{11} versus C2, for a tapped capacitor resonator. The inductor is 10nH, with a Q $=18$

PA can be employed to increase efficiency. To avoid wasting power, the active element(s) in the PA should switch on and off completely and have close to 0V across them when strongly conducting - implying the PA should be driven at or near its maximum possible voltage swing. Hence, the most efficient output power for the PA is determined by the real part of its load impedance R_{Load} and available zero-to-peak voltage swing $v_{o,max}$.

$$P_{maxeff} = \frac{v_{o,max}^2}{2R_{Load}} \qquad (8.28)$$

To design an efficient PA with very low power output, then, we need a small $v_{o,max}$ and large R_{Load}.

PA Topology and V_{max}

Generally speaking, supply voltage is fixed by other design constraints, so $v_{o,max}$ can only be reduced by changing the PA topology. Figure 8.13, illustrates a few different PA topologies with different values for $V_{o,max}$.

At the far right, two identical push-pull PAs are effectively stacked on top of one another to cut $V_{o,max}$ by a factor of four. Thus, for a given average supply current Idc, this PA can deliver 4 times as much current to the load as the PA at the far left; each electron in the output current flows through the load four times. The stacked push-pull topology was presented in [9],

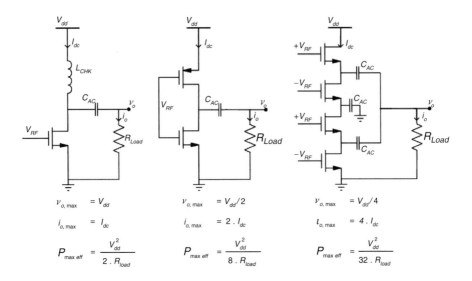

Fig. 8.13. Various PA topologies maintain efficiency over a wide range of output powers without resonant networks or changing supply voltage

demonstrating 40% efficiency with $250\mu W$ output power in the 900MHz ISM band.

Boosting Load Impedance with Resonant Networks

Another key to increasing efficiency at low power output is to increase R_{Load}. In narrowband systems, R_{Load} can be varied over a very wide range by using a resonant LC matching network to transform the original antenna impedance R_A. The ratio of the transformed R_L to the original RA typically scales with the square of the overall network quality factor (Q_{net}). Hence, transforming impedance by large ratios is only useful for narrowband systems where moderate values of Q are acceptable. The parasitic series resistance of real inductors and capacitors will place an upper limit on Q_{net} and, therefore, the maximum achievable impedance transformation ratio. The network will tend to have poor efficiency as it approaches that maximum transformation ratio. As an example, we compute the output impedance and network efficiency for the tapped-capacitor described in Section 8.3.3 as a function of R_A and Q_L. To simplify the calculations, we assume the inductor is much lower Q than the capacitor Q (see Section 8.3.3) and ignore the series resistance of the capacitors. The output impedance of the network is expressed in 8.23 and the efficiency is shown below, with $Q_{C,eff}$.

$$E = \frac{Q_L}{Q_C + Q_L} \tag{8.29}$$

The highest efficiency occurs at $C_2 = 0$ where $Q_{C,eff}$ is minimized. Note that this is actually the inverse of 8.26, which represents the network noise factor. As the impedance ratio is increased, a larger proportion of the signal power is dissipated in R_L, resulting in lower efficiency.

8.4 A Low Energy 2.4GHz Transceiver

In this section, we examine a low energy transceiver with respect to the energy saving techniques discussed thus far. A much more detailed, circuit focused analysis of the system is carried out in [73]. This 2.4GHz transceiver, implemented in a $0.13\mu m$ RF CMOS process, achieves 1nJ per received bit and 3nJ per transmitted bit with $300\mu m$ transmit power and 7dB receiver noise figure. The resulting energy efficiency figure of merit η is -30dB, and is actually dominated by the high $\frac{E_b}{N_0}$ required to demodulate 2-FSK noncoherently. The transceiver block diagram is shown in Figure 8.14.

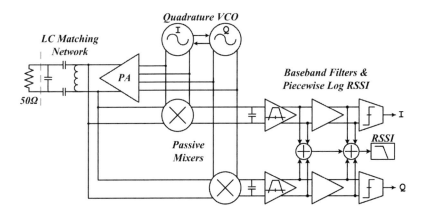

Fig. 8.14. Transceiver block diagram

A 400mV supply was chosen for this system to accommodate a single solar cell as the power source. In sunlight the entire transceiver could operate continuously from a 2.6mm x 2.6mm silicon solar cell [74]. Because of the reduced supply voltage, all circuits are made differential to increase available swing. This transceiver uses 2-FSK with a relatively large tone separation, effectively trading spectral efficiency for a simplified low-power architecture. This tradeoff is particularly favorable for sensor network applications, wherein data rates below 1Mbps are the norm and 85MHz of unlicensed spectrum is available in the 2.4GHz ISM band.

In the transmitter, a cross-coupled LC VCO directly drives an efficient nonlinear PA - taking advantage of the relaxed phase accuracy requirements to eliminate upconversion mixers and transmit buffers by using direct VCO modulation. Furthermore, the PA input capacitance is incorporated into the resonant LC tank of the VCO to minimize the current consumed driving the PA. Since the supply voltage is so low, the NMOS-only VCO architecture was chosen to achieve maximize swing. The differential PA drives a tapped-capacitor resonator to boost its load impedance from 50Ω to about $1k\Omega$. The PA achieves 45% efficiency at $300\mu W$ output power and the overall transmitter efficiency is 30%. A plot of transmitted power versus total power consumed is shown in Figure 8.15.

The tapped capacitor network at the PA output is also used in the receiver front-end to achieve impedance matching and substantial voltage gain - effectively supplanting an RF LNA. This network interfaces directly to highly linear CMOS passive mixers designed to present a high impedance to the LC network to avoid reducing its voltage gain. A reconfigurable front-end was devised to limit capacitive loading on the LC network and the VCO by reducing transistor count in the front-end. In essence, a single quad of transistors can be configured either as a PA or as a mixer, depending on bias voltages and the states of a couple switches. Figure 8.16 illustrates this reconfigurable topology.

The passive mixers downconvert the desired signal to baseband and attenuate wideband interference with a 1MHz first order low-pass filter at the output. The overall voltage gain from the balanced 50Ω receiver input to the mixer output is about 17dB. The mixer outputs drive a sequence of bandpass filters with enough gain to convert the incoming signal to a 1-bit quantized waveform for simple demodulation. The cascaded receiver noise factor versus power consumption is shown in Figure 8.15, and a summary of the measured performance is in Figure 8.17.

8.5 Summary and Conclusions

We began this chapter using Shannons capacity theorem to determine the fundamental energy requirements of wireless communication from the most general, theoretical standpoint. As a natural consequence of this discussion, we proposed a figure of merit η for evaluating the energy efficiency of real wireless systems relative to fundamental limits. Next, we separated the major contributors to low efficiency by expressing this figure of merit in terms of a practical systems $(\frac{E_b}{N_0})_{min}$, global transmit efficiency, and receiver noise factor. Examining $(\frac{E_b}{N_0})_{min}$ figures and architectural implications of some popular modulation schemes revealed that modulation is extremely important in determining a systems energy efficiency and suggested that simple schemes such as OOK, 2-FSK, and PSK would be well suited to low-energy systems. Furthermore, error correcting codes (ECC) with substantial coding gain are

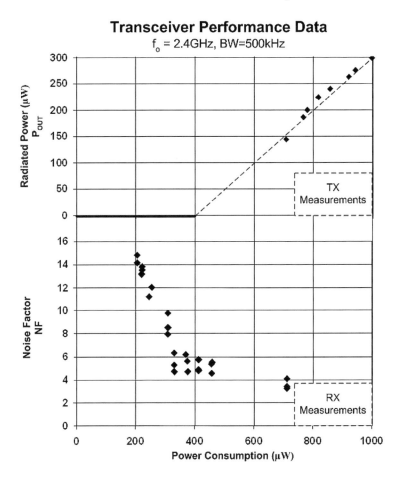

Fig. 8.15. Measured transceiver performance data. This 2.4GHz radio operates from a 400mV supply and achieves 4nJ/bit communication with 92dB link margin. PA efficiency is 44% and the power overhead is estimated as: $P_{OH,TX} = 400\mu W$ and $P_{OH,RX} = 170\mu W$

becoming feasible even for low power systems, owing to the continued scaling down of digital CMOS [60]. With the assumption of an OOK, FSK, or PSK system, several circuit techniques were presented aimed at boosting PA efficiency, reducing power overhead, or achieving low noise factor with a low power receiver. Finally, we looked at a 2.4GHz transceiver incorporating many

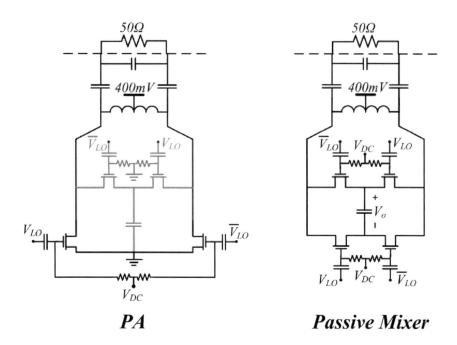

Fig. 8.16. Reconfigurable PA/Mixer topology. Sharing transistors of the PA and mixer substantially reduces parasitic loading on the input LC network and the VCO tank

of these techniques to achieve high energy efficiency at very low transmit power.

Overall		Value
Supply Voltage	Min/Typ/Max	360/400/600 mV
2-FSK Deviation	Min/Max	300/1000 kHz
RX		
Power Consumption	Min/Max	200/750 µW
Noise Figure	Min/Max	5.1/11.8 dB
IIP3	Typ.	-7.5 dBm
TX		
Power Consumption	Min/Max	700/1120 µW
Output Power	Min/Max	140/320 µW
PA Efficiency	$200 < P_{OUT} < 300$ µW	>44 %
VCO		
Power Consumption	$\text{Min}_{\text{I Only}}/\text{Max}_{\text{I+Q}}$	160/700 µW
Frequency	Min/Max	1.95/2.38 GHz
Quadrature Mismatch	Meas. $\Delta\Phi$ at I&Q Mixer Out	$|\Delta\Phi - 90| < 5$ deg
Phase Noise @1MHz	@ P_{VCO} = 270 µW	-106 dBc/Hz

Fig. 8.17. Summary of measured performance

9

CONCLUSIONS

The field of wireless sensing holds the potential to greatly change the way we live. Pushing these devices to smaller, cheaper, and more functional implementations greatly increases the scope of their application. However, there are huge technological challenges in making this visions a reality. This book has discussed technologies and methodologies for overcoming these obstacles. Input from multiple research groups was compiled to yield a diverse glimpse into the field of ultra-low power transceiver design. The following list highlights a few of the contributions of this book.

- The design and integration of energy harvesting electronics
- Ultra-low power integrated circuit/MEMS co-design techniques
- The presentation of a sub-mW super-regenerative transceiver architecture
- A detailed description of a low power CMOS ultra-wideband transceiver for sensor networks
- Novel methods of modular packaging for sensor nodes
- A novel low voltage, low power CMOS narrowband RF transceiver

Implementation of the RF transceiver is one of the greatest challenges in the realization of energy scavenging sensor networks. The transceiver must consume very little power (less than 1mW), be fully integrated, and exhibit a very rapid start-up time. These requirements are in direct conflict with each other. To overcome these tradeoffs, emerging RF MEMS technologies were investigated. These devices are shown to greatly reduce the complexity and power consumption of low datarate transceivers.

An RF MEMS/CMOS co-design methodology was developed that allows the simultaneous optimization of the CMOS active devices and the MEMS components. Multiple proof-of-concept circuit blocks were developed and tested to verify these concepts. These include a single-ended Pierce oscillator and a new differential oscillator topology. With these devices, it was possible to demonstrate a self-contained energy scavenging transmit beacon. This device continues to operate on scavenged energy after over three years of operation.

These proposed methodologies built the foundation for the next step: designing a wireless link based on these principles. An architecture exploration was performed, and two prototype transceivers were implemented and tested. The first was a tuned RF architecture. This architecture takes advantage of the array capability of RF MEMS devices, and is scalable to multiple discrete channels. The second was a super-regenerative receiver. This receiver achieves well below state-of-the-art power consumption by employing MEMS resonators and subthreshold CMOS design techniques. A fully integrated, (1×2)mm^2 transceiver was developed. For both transceivers, an indoor 20m wireless link was demonstrated to verify robust operation. These transceivers can be directly applied to various wireless sensing scenarios. The methodologies presented in this book can be applied to the design of other communications systems. As an example, the design of a communication link for an ultra-dense sensor network $(>10\frac{nodes}{m^2})$ would greatly benefit from these techniques. As RF MEMS technologies mature, they will be more reliable, more fully integrated, and handle more power. The maturity of these technologies will open up a new dimension of circuit design.

The benefits of MEMS/CMOS co-design techniques were shown to impact other areas of the sensor node besides the transceiver. For example, these devices can be used to replace quartz crystals as the system time base. This is advantageous because quartz crystals are bulky and available only with discrete frequency and impedance values. A design example was shown that performed a co-design a 16MHz electrostatic resonator with custom CMOS oscillators, greatly reducing the form factor of the system reference clocks.

A contributed chapter by Julien Ryckaert and Steven Sanders of IMEC, Belgium presented an ultra-low power, ultra wideband (UWB) transceiver implementation. This work demonstrates that low power transceivers for sensor networks are not necessarily constrained to the narrowband category. Systematically, the authors build up the transceiver requirements beginning with the allotted spectrum allocation. Novel methods of pulse-generation are presented, and detailed circuit design techniques are demonstrated.

Additionally, this work provides groundbreaking philosophies on modular packaging for sensor nodes. This strategy involves dedicated boards for different sensor node functionality. For example, there is a dedicated communications board, computation board, and a power management/sensor board. In order to achieve the 3D SiP form factor, a novel technique for layer interconnection was applied. The different functional layers (communication, computation, application) were designed on different boards and consequently stacked on each other through a dual row of fine pitch solder balls.

The next chapter was contributed by Ben Cook and Prof. Kris Pister from the University of California, Berkeley. This chapter presents an ultra-low voltage, low power narrowband transceiver. This devices utilizes an extremely low (400mV) supply voltage to reduce the power consumption. In addition, to low noise amplifier (LNA) is used, improving the linearity significantly and reducing the power consumption of the receiver front-end. The authors provide detailed circuit design techniques as well as measured results from a fully integrated transceiver implementation.

Although the techniques in this book were designed with an eye towards wireless sensor networks, they can certainly be extended to other circuits and systems in many application spaces. In particular, the emerging field of ubiquitous health monitoring and autonomous implantable electronics can leverage these techniques to integrate very low power communications with diminishing form-factors.

References

1. J. Whitehead, *Super-Regenerative Receivers*. Cambridge Press, 1950.
2. B. Cook, A. Molnar, and K. Pister, "Low power RF design for sensor networks," *Proceedings of the IEEE Radio Frequency Integrated Circuits (RFIC) Conference*, 2005.
3. S. Roundy, *Energy scavenging for wireless sensor nodes with a focus on vibration to electricity conversion*. PhD thesis, University of California, Berkeley, May 2003.
4. J. Rabaey, J. Ammer, T. Karalar, S. Li, B. Otis, M. Sheets, and T. Tuan, "Picoradios for wireless sensor networks: The next challenge in ultra-low power design," *IEEE ISSCC Digest of Technical Papers*, pp. 200–01, Feb 2002.
5. J. Kao, S. Narendra, and A. Chandrakasan, "Subthreshold leakage modeling and reduction techniques," *Proceedings of IEEE/ACM International Conference on Computer Aided Design*, pp. 141–148, 2002.
6. L. Zhou, J. Kahn, and K. Pister, "Corner-cube retroreflectors based on structure-assisted assembly for free-space optical communcation," *IEEE J. Microelectromechanical Systems*, vol. 12, pp. 233–242, June 2003.
7. A.-S. Porret, T. Melly, D. Python, C. Enz, and E. Vittoz, "An ultralow-power UHF transceiver integrated in a standard digital CMOS process: architecture and receiver," *IEEE J. Solid-State Circuits*, vol. 36, pp. 452–66, Mar 2001.
8. V. Peiris *et al.*, "A 1V 433/868MHz 25kb/s-FSK 2kb/s-OOK RF Transceiver SoC in standard digital 0.18μm CMOS," *IEEE ISSCC Digest of Technical Papers*, p. 258, Feb 2005.
9. A. Molnar, B. Lu, S. Lanzisera, B. Cook, and K. Pister, "An ultra-low power 900 MHz RF transceiver for wireless sensor networks," *IEEE Custom Integrated Circuits Conference*, 2004.
10. E. Vittoz and J. Fellrath, "CMOS analog integrated circuits based on weak-inversion operation," *IEEE J. Solid State Circuits*, vol. 12, pp. 224–231, June 1977.
11. N. Pletcher, "Micro Power Radio Frequency Oscillator Design," Master's thesis, University of California, Berkeley, Dec 2004.
12. B. Bircumshaw *et al.*, "The radial bulk annular resonator: Towards a 50ohm RF MEMS filter," *12th Int. Conf. on Solid State Sensors, Actuators and Microsystems*, vol. 35, pp. 875–878, June 2003.

13. K. Lakin, "Thin film resonators and high frequency filters," *http://www.tfrtech.com*, June 2001.

14. B. Otis and J. Rabaey, "A 300μw 1.9GHz CMOS oscillator utilizing micromachined resonators," *IEEE J. Solid State Circuits*, July 2002.

15. S. Sheng and R. Brodersen, *Low-Power CMOS Wireless Communications*. Kluwer, 1998.

16. J. Smith, "High density, low parasitic direct integration by fluidic self-assembly (FSA)," *Dig. IEEE International Electron Devices Meeting*, pp. 201–204, 2001.

17. R. Ruby, P. Bradley, J. L. III, Y. Oshmyansky, and D. Figueredo, "Ultraminiature high-Q filters and duplexers using FBAR technology," *IEEE ISSCC Digest of Technical Papers*, pp. 120–21, Feb 2001.

18. M. Aissi, E. Tournier, M. Dubois, G. Parat, and R. Plana, "A 5.4 ghz 0.35 μm BiCMOS FBAR resonator oscillator in above-IC technology," in *ISSCC Digest of Technical Papers*, Feb. 2006.

19. B. Otis, "The Design and Implementation of an Ultra Low Power RF Oscillator using Micromachined Resonators," Master's thesis, University of California, Berkeley, Dec 2002.

20. G. Chien and P. Gray, "A 900-MHz local oscillator using a DLL-based frequency multiplier technique for PCS applications," *IEEE J. Solid-State Circuits*, vol. 35, pp. 1996–99, Dec 2000.

21. Q. Huang and P. Basedau, "Design considerations for high-frequency crystal oscillators digitally trimmable to sub-ppm accuracy," *IEEE Trans. on VLSI Systems*, vol. 5, pp. 408–16, Dec 1997.

22. E. Vittoz, M. Degrauwe, and S. Bitz, "High-performance crystal oscillator circuits: theory and application," *IEEE J. Solid State Circuits*, vol. 23, pp. 774–83, June 1988.

23. K. Kwok and H. Luong, "Ultra-low-voltage high-performance CMOS VCOs using transformer feedback," *IEEE Journal of Solid-State Circuits*, vol. 40, pp. 652–660, 3 2005.

24. N. Pletcher and J. Rabaey, "A 100μW, 1.9GHz oscillator with fully-digital frequency tuning," *Submitted: Proceedings of the IEEE ESSCIRC*, 2005.

25. D. Ruffieux, "A high-stability, ultra-low-power differential oscillator circuit for demanding radio applications," *Proceedings of the IEEE ESSCIRC*, pp. 85–88, Oct 2002.

26. J. Larson and Y. Oshmyansky, "Measurement of effective k_t^2, Q, Rp, Rs vs. temperature for Mo/AlN FBAR resonators," *IEEE Ultrasonics Symposium*, 2002.

27. S. Roundy, B. Otis, Y. Chee, J. Rabaey, and P. Wright, "A 1.9GHz RF transmit beacon using environmentally scavenged energy," *Proceedings of the ISLPED*, Sept 2003.

28. B. Otis, Y. Chee, R. Lu, N. Pletcher, and J. Rabaey, "An ultra-low power MEMS-based two-channel transceiver for wireless sensor networks," *Proceedings of the IEEE VLSI Symposium*, June 2004.

29. R. Lu, "CMOS low noise amplifier design for wireless sensor networks," Master's thesis, University of California, Berkeley, Dec 2003.

30. R. Meyer, "Low-power monolithic RF peak detector analysis," *IEEE J. Solid State Circuits*, vol. 31, Jan 1995.

31. F. Assaderaghi, D. Sinitsky, S. Parke, J. Bokor, P. Ko, and C. Hu, "Dynamic threshold-voltage MOSFET (DTMOS) for ultra-low voltage VLSI," *IEEE Trans. Elec. Devices*, vol. 44, Mar 1997.

32. E. Armstrong, "Some recent developments of regenerative circuits," *Proc. IRE*, vol. 10, pp. 244–260, Aug 1922.

33. N. Joehl, C. Dehollain, P. Favre, P. Deval, and M. Declercq, "A low-power 1-GHz super-regenerative transceiver with time-shared PLL control," *IEEE J. Solid State Circuits*, vol. 36, pp. 1025–1031, July 2001.

34. A. Vouilloz *et al.*, "A low-power CMOS super-regenerative receiver at 1GHz," *IEEE J. Solid-State Circuits*, vol. 36, pp. 440–451, Mar 2001.

35. J. Chen, F. M., and H. J., "Fully integrated auto-calibrated super-regenerative receiver," in *ISSCC Digest of Technical Papers*, Feb. 2006.

36. B. Otis, Y. Chee, and J. Rabaey, "A 400μW Rx, 1.6mW Tx super-regenerative transceiver for wireless sensor networks," *IEEE ISSCC Digest of Technical Papers*, p. 396, Feb 2005.

37. A. Zverev, *Handbook of filter synthesis*. Wiley, 1967.

38. F. Moncunill, O. Mas, and P. Pala, "A direct-sequence spread-spectrum super-regenerative receiver," *IEEE International Symposium on Circuits and Systems (ISCAS)*, pp. I–68–I–71, May 2000.

39. A. Franke, T.-J. King, and R. Howe, "Post-CMOS modular integration of poly-SiGe microstructures using poly-Ge sacrificial layers," *Solid-State Sensor and Actuator Workshop Technical Digest*, pp. 18–21, June 2000.

40. H. Takeuchi, E. Quevy, S. Bhave, T.-J. King, and R. Howe, "Ge-blade dama-scene process for post-CMOS integration of nano-mechanical resonators," *IEEE Electron Device Letters*, 2004.

41. C. Nguyen and R. Howe, "An integrated CMOS micromechanical resonator high-Q oscillator," *IEEE J. Solid State Circuits*, vol. 34, pp. 440–455, Apr 1999.

42. R. Ruby, P. Bradley, and J. L. III, "FBAR microcap technology," *IEEE ISSCC Digest of Technical Papers*, Feb 2002.

43. J. Rabaey, M. Ammer, J. da Silva, J.L., D. Patel, and S. Roundy, "Picoradio supports ad hoc ultra-low power wireless networking," *IEEE Computer*, vol. 33, pp. 42–48, July 2000.

44. http://www.bluetooth.com , http://www.zigbee.org.

45. M. Win and R. Scholtz, "Impulse radio: how it works," in *IEEE Communications Letters*, pp. 36–38, Feb. 1998.

46. F. C. C. (FCC), "Revision of part 15 regarding ultra-wideband transmission systems." First Report and Order, ET Docket, 98–153, FCC 02-48, adopted Feb. 2002, released Apr. 2002, available at http://www.fcc.gov.

47. "standard draft proposal." IEEE 802.15.4a www.ieee802.org/15/pub/TG4a.html.

48. IEEE 802.15.3a www.ieee802.org/15/pub/TG3a.html.

49. H. Sheng, P. Orlik, A. Haimovitch, L. Cimini, and J. Zhang, "On the spectral and power requirements for ultra-wideband transmission," in *IEEE Proc. of Int. Conf. on Communications*, vol. 1, pp. 738–742, May 2003.

50. J. Han and C. Nguyen, "A new ultra-wideband, ultra-short monocycle pulse generator with reduced ringing," in *IEEE Microwave and Wireless Components Letters*, pp. 206–208, 2002.

51. K. Marsden, H. Lee, D. Ha, and H. Lee, "Low power CMOS re-programmable pulse generator for UWB systems," in *IEEE Conference on UWB Systems and Technologies*, pp. 443–447, 2003.

52. Y. Choi, "Gated UWB pulse signal generation," in *IEEE Joint International Workshop of UWBST and IWUWBS*, pp. 122–124, May 2004.

53. J. Ryckaert, C. Desset, A. Fort, M. Badaroglu, V. De Heyn, P. Wambacq, G. Van der Plas, S. Donnay, B. Van Poucke, and B. Gyselinckx, "Ultra-wide band transmitter for low-power wireless body area networks: Design and evaluation," *IEEE Transactions on Circuits and Systems*, vol. 52, Dec. 2005.

54. J. Ryckaert, M. Badaroglu, V. De Heyn, G. Van der Plas, P. Nuzzo, A. Baschirotto, S. D'Amico, C. Desset, H. Suys, M. Libois, B. Van Poucke, P. Wambacq, and B. Gyselinckx, "A 16mA UWB 3-to-5Ghz 20Mpulses/s quadrature analog correlation receiver in 0.18μm CMOS," in *ISSCC Digest of Technical Papers*, pp. 114–115, Feb. 2006.

55. J. Ryckaert, M. Badaroglu, C. Desset, V. de Heyn, G. Van der Plas, P. Wambacq, B. Van Poucke, and S. Donnay, "Carrier-based uwb impulse radio: Simplicity, flexibility, and pulser implementation in 0.18μm cmos," *International Conference on Ultrawideband, ICU 2005*, 2005.

56. R. Blazquez, F. Lee, D. Wentzloff, B. Ginsburg, J. Powell, and A. Chandrakasan, "Direct conversion pulsed uwb transceiver architecture," *Proc. of Design, Automation and Test in Europe*, Mar. 2005.

57. S. D'Amico, T. Grassi, J. Ryckaert, and A. Baschirotto, "A up to 1Ghz low-power continuous-time 3rd-order filter+integrator chain for wireless body-area network receivers," in *IEEE Prime*, 2005.

58. G. Van der Plas, S. Decoutere, and S. Donnay, "A 0.16pJ/conversion-step 2.5mW 1.25GS/s 4b ADC in a 90nm CMOS process," in *ISSCC Digest of Technical Papers*, pp. 566–567, Feb. 2006.

59. J. Proakis, *Digital Communications, 4th ed.* McGraw-Hill, 2001.

60. C. Desset and A. Fort, "Selection of channel coding for low-power wireless systems," *Proc. of the IEEE Vehicular Technology Conference*, 2003.

61. D. Magill, F. Natali, and G. Edwards, "Spread-spectrum technology for commercial applications," *Proceedings of the IEEE*, vol. 82, pp. 572–584, 1994.

62. H. Hashemi, "The indoor radio propagation channel," *Proc. of the IEEE*, vol. 81, pp. 943–968, 1993.

63. W. Kluge *et al.*, "A fully integrated 2.4ghz IEEE 802.15.4 compliant transceiver for ZigBee applications," *IEEE ISSCC Digest of Technical Papers*, Feb 2006.

64. B. Cook, A. Berny, A. Molnar, S. Lanzisera, and K. Pister, "An ultra-low power 2.4ghz rf transceiver for wireless sensor networks in 130nm CMOS with 400mV supply and an integrated passive RX front-end," *IEEE ISSCC Digest of Technical Papers*, Feb 2006.

65. P. Choi, "An experimental coin-sized radio for extremely low-power WPAN (IEEE 802.15.4) application at 2.4 GHz," *IEEE Journal of Solid-State Circuits*, vol. 38, pp. 2258–2268, 2003.

66. ChipCon, "CC1010 Datasheet," *http://www.chipcon.com*.

67. ChipCon, "CC2420 Datasheet," *http://www.chipcon.com*.

68. A. Communications, "Power consumption and energy efficiency comparison of WLAN products," *http://www.atheros.com/pt/papers*.

69. M. H. Perrott, T. L. I. Tewksbury, and C. G. Sodini

70. R. B. Staszewski *et al.*, "All-digital TX frequency synthesizer and discrete-time receiver for Bluetooth radio in 130nm CMOS," *IEEE Journal of Solid-State Circuits*, vol. 39, 2004.

71. E. Hegazi and A. Abidi, "A 17 mW transmitter and frequency synthesizer for 900 MHz GSM fully integrated in 0.35um CMOS," *IEEE Journal of Solid-State Circuits*, vol. 38, 2003.

72. A. Berny, "Dissertation: Analysis and design of wideband LC VCOs," *Department of Electrical Engineering: University of California-Berkeley*, 2006.
73. B. Cook, A. Berny, A. Molnar, S. Lanzisera, and K. Pister, "Low power 2.4ghz rf transceiver with passive RX front-end and 400mV supply," *IEEE Journal of Solid-State Circuits*, Jan 2006.
74. Solarbotics, "OSRAM BPW 34 Datasheet," *http://www.solarbotics.com*.

Index

V

Voltage regulator, 36

W

Weak inversion CMOS, 13, 43, 47, 53, 72
Wideband signaling, 117

Printed in the United States